Principles and Procedures of Multiple Matrix Sampling

Principles and Procedures of Multiple Matrix Sampling

David M. Shoemaker
Southwest Regional Laboratory for Educational Research and Development

Ballinger Publishing Company • Cambridge, Mass.
A Subsidiary of J.B. Lippincott Company

This work was developed under a contract with the U.S. Office of Education and the National Institute of Education of the Department of Health, Education and Welfare. However, the content does not necessarily reflect the position or policy of that Agency, and no official endorsement of these materials should be inferred.

Library of Congress Catalog Card Number: 73–9577

International Standard Book Number: 0–88410–153–3

Printed in the United States of America

Library of Congress Cataloging in Publication Data

Shoemaker, David M 1940–
Principles and procedures of multiple matrix sampling.
Bibliography: p.
1. Sampling (Statistics) 2. Matrices. I. Title.
QA276.6.S48 519.5'2 73–9577
ISBN 0–88410–153–3

Contents

List of Figures

List of Tables

Preface

Descriptions of multiple matrix sampling and explorations into areas of application are scattered across a variety of technical journals. There is no single book or monograph which describes, structures, and unifies all of this material. Yet there is a need for such a document both as a reference source and as a textbook. Additionally, such a compilation should facilitate the growth of knowledge about multiple matrix sampling by permitting contemporary research to start where that of others has left off.

The original impetus for most new ideas is often difficult to pinpoint exactly and such is the case with multiple matrix sampling. Although William E. Turnbull, Robert L. Ebel, and Frederic M. Lord while working at the Educational Testing Service undoubtedly discussed in the early 1950s the possibility of such a technique for estimating the normative distribution for standardized tests, it was Lord, Robert Hooke, and John Tukey (the latter two at Princeton University) who subsequently developed the statistical foundation for the majority of estimation procedures involved in multiple matrix sampling. The first consolidation of research on this measurement procedure appeared in 1968 in Chapter 11 of *Statistical Theories of Mental Test Scores*, a book coauthored by Lord and Melvin R. Novick. Although the chapter is sound statistically, it does not encompass the degree of detail and depth of explanation required by the majority of practitioners who desire to implement this research procedure in a particular investigation. This book is designed both to remedy this situation and to supplement the content in terms of recent developments.

Throughout this book an attempt has been made to keep the practitioner clearly in mind. The emphasis is clearly on the *why*, *when*, and *how* to use multiple matrix sampling. The topics covered range from a description of multiple matrix sampling to use of the computer program in analyzing the data. All discussions and guidelines contained in the book reflect theoretical and empirical results reported in the literature as well as personal experiences of the author in implementing multiple matrix sampling in a variety of applied situations.

Following these introductory remarks, the book is subdivided into seven chapters, two appendixes, and a bibliography. In Chapter 1, multiple matrix sampling is defined and contrasted with traditional measurement models; additionally, the advantages, limitations, and potential applications of the procedure are considered in some detail. Guidelines for implementing the procedure in a project are discussed in Chapter 2, and in Chapter 3 the relevant equations are listed and applied to a sample data set. Chapter 4 describes a computer model developed by the author for simulating multiple matrix sampling. Although the model may be used in numerous ways, it is included primarily as a teaching device to aid in the understanding of multiple matrix sampling. The relationship between multiple matrix sampling and statistical hypothesis testing procedures is discussed in Chapter 5. Several applications of multiple matrix sampling are discussed in detail in Chapter 6. Some prospects for the future discussed in Chapter 7, and that found in the Bibliography should come as a surprise to none. Because multiple matrix sampling is a procedure in which computer data processing is essential and not ornamental, a computer program is required. One designed specifically by the author for this purpose is given in Appendix A. The computer program discussed in Chapter 4 is given in Appendix B.

Acknowledgments

I would like to thank Frederic M. Lord of the Educational Testing Service, Richard E. Schutz and Robert Baker of the Southwest Regional Laboratory for Educational Research and Development, Ken Sirotnik of the University of California at Los Angeles, and Thomas R. Knapp of the University of Rochester for reading preliminary drafts of this book and contributing several improvements. Any blunders, awkward passages, or inaccuracies that remain cannot be laid at their doorsteps. Appreciated additionally is the assistance given me by my associate Tej Pandey.

Summary of Notations

The symbols used commonly in this book are identified below. Notation used at only one point will not appear in this summary.

K	Number of items in universe
N	Number of examinees in population
t	Number of subtests
k	Number of items in each subtest
n	Number of examinees to which each subtest is administered
$(t/k/n)$	An ordered triad denoting a specific sampling plan in multiple matrix sampling
μ_1'	Mean universe score
μ_2	Variance of universe scores (second central moment)
μ_3	Third central moment of universe scores
μ_4	Fourth central moment of universe scores
	An estimate of
$\hat{\mu}'_{1_s}$	An estimate of μ_1' using results from subtest s
$\hat{\mu}_{2_s}$	An estimate of μ_2 using results from subtest s
$\hat{\mu}_{3_s}$	An estimate of μ_3 using results from subtest s
$\hat{\mu}_{4_s}$	An estimate of μ_4 using results from subtest s
$\hat{\mu}'_{1_p}$	A pooled estimate of μ_1'
$\hat{\mu}_{2_p}$	A pooled estimate of μ_2
$\hat{\mu}_{3_p}$	A pooled estimate of μ_3

$\hat{\mu}_{4_p}$	A pooled estimate of μ_4
σ_I^2	Component of variance for items
σ_E^2	Component of variance for examinees
σ_{IE}^2	Component of variance for item-by-examinee interaction
$\hat{\sigma}_{I_s}^2$	An estimate of σ_I^2 using results of subtest s
$\hat{\sigma}_{I_p}^2$	A pooled estimate of σ_I^2
VAR(X)	Variance of X
SD(X)	Standard deviation of X
SE(X)	Standard error of estimate for X
MS	Mean squared deviation
EMS	Expected mean squared deviation
SS	Sum of squared deviations
df	Degrees of freedom
KR_{20}	Kuder-Richardson Formula 20 coefficient of reliability

Principles and Procedures of Multiple Matrix Sampling

Chapter One

Characteristics, Advantages, and Applications of Multiple Matrix Sampling

The majority of contemporary psychometric procedures reflect strongly the original impetus of the psychometric movement, that is, the measurement of individual differences. Historically, individual differences have been investigated, and appropriately so, using the *matched-items* model in which a single set of test items is administered in a standardized procedure to all, or a sample, of the examinee population under consideration. Exemplars of such methodology are numerous and range from the "mental tests" devised by such pioneers as James M. Cattell, Sir Francis Galton, and Alfred Binet to the achievement tests used today in almost every classroom. In each instance, the goal of the measurement procedure was and is the same: assessing the degree to which each individual person can perform a specified task.

Understandably so and undoubtedly for lack of a reasonable alternative, procedures appropriate for the assessment of *individual* differences have been transferred completely to investigations concerned primarily with the measurement of *group* differences. An example of the latter is an investigation designed to assess the relative merits of two uniquely different reading programs constructed for kindergarten students. In this case, the researcher is interested primarily in contrasting the performance of the *group* of students participating in the first program with the performance of the *group* of students participating in the second. Differences in performance among individual students are of little or no importance. Of major concern is determining the performance level of a *group* of students. The point to be made is simply this: the methodology employed successfully in the assessment of individual differences is neither the appropriate nor the most efficient methodology for group assessment. Multiple matrix sampling has been demonstrated theoretically and empirically to be the preferred procedure for group assessment and one superior to the matched-items model. This conclusion is of no minor import because quite often an investigator is concerned solely with assessing group performance.

MULTIPLE MATRIX SAMPLING MODEL

The matched-items model and the multiple matrix sampling model are contrasted readily by considering the data base which would be generated if each individual in the population of N examinees were administered each item comprising the K-item universe. (Throughout this book, the term *population* will always be used to refer to the total group of examinees whose performance is being assessed and *universe*, to the total number of items on which information is desired.) Such a data base is illustrated in Figure 1-1 and the arrangement is referred to commonly as an examinee-by-item matrix. Test items are frequently

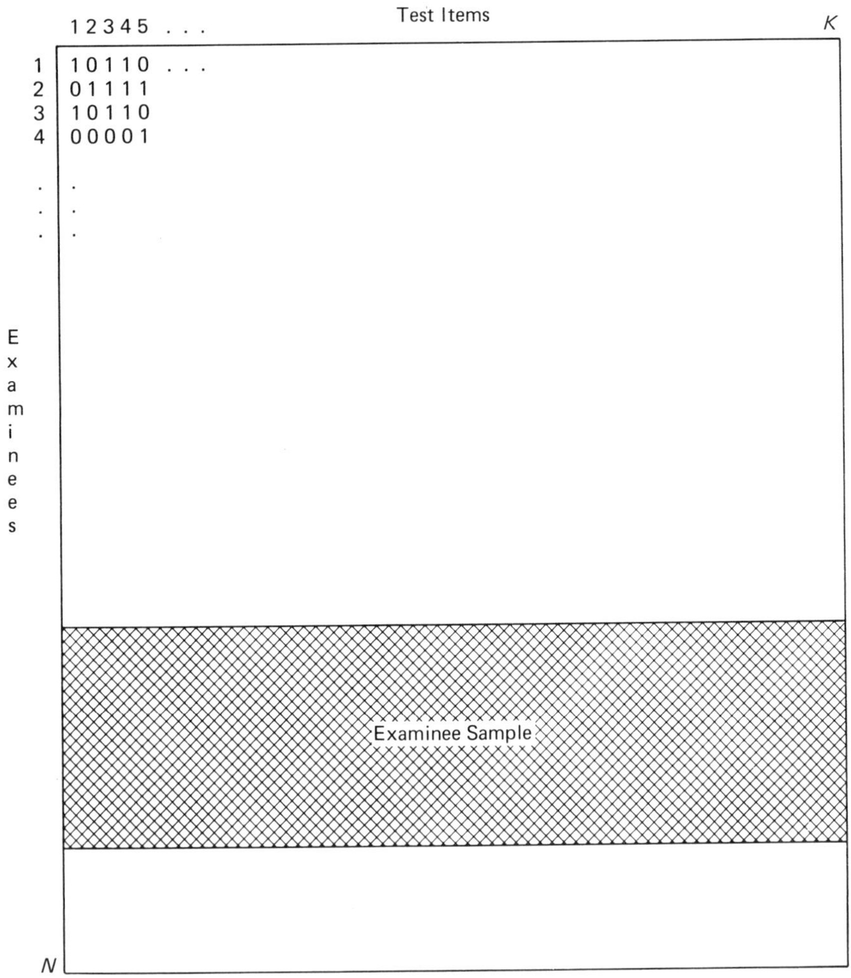

Figure 1-1. Examinee-by-Item Matrix Illustrating Examinee-Sampling

scored dichotomously (1 = pass, 0 = fail) and such is the case in Figure 1-1, where, for example, examinee 1 passed item 1, failed item 2, passed items 3 and 4, and failed item 5.

Within the framework of the matrix depicted in Figure 1-1, the matched-items model used in the assessment of individual differences is referred to as *examinee-sampling* because all items in the universe are administered to a subgroup of examinees sampled from the population. For example, if the universe of interest consisted of 300 items, each examinee selected would be administered all of the 300 items. The cross-hatched area in Figure 1-1 illustrates this sampling procedure. By contrast, multiple matrix sampling involves the joint sampling of both items *and* examinees, as is indicated in Figure 1-2. Data from

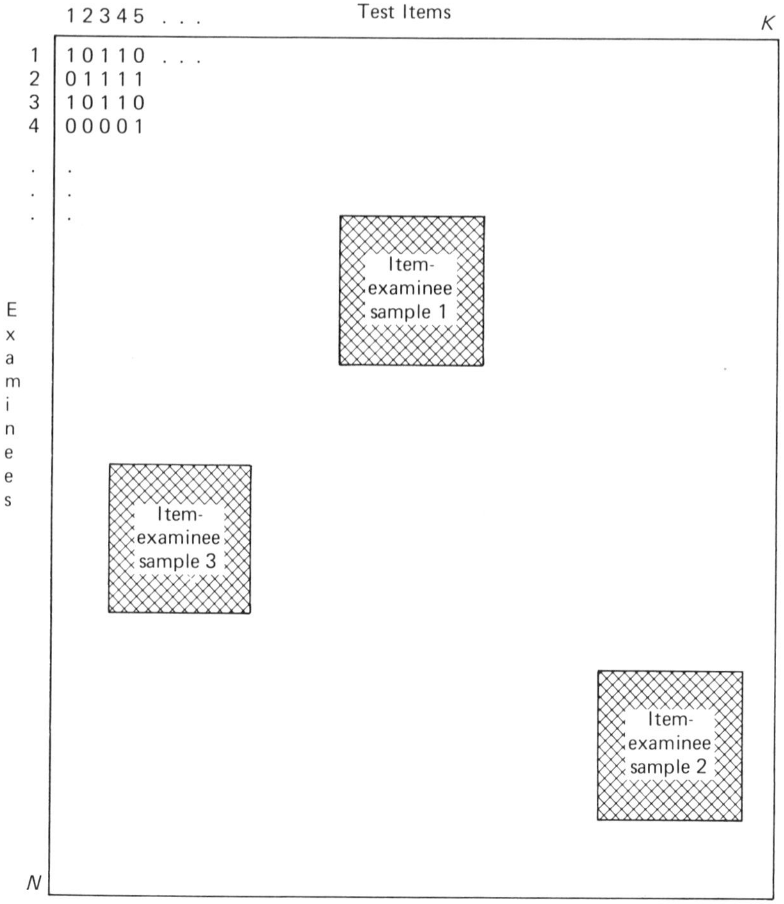

Figure 1-2. Examinee-by-Item Matrix Illustrating Multiple Matrix Sampling

item-examinee sample 1 were obtained by administering a subset of items selected from the universe to a subgroup of examinees selected from the population. Replicating this procedure produces item-examinee samples 2 and 3 and suggests concomitantly the derivation of the expression "multiple matrix sampling." Using the 300-item universe as an example, the first examinee subgroup could, in multiple matrix sampling, be administered items 1 to 100; the second, items 101 to 200, and; the third, items 201 to 300. (There are numerous ways in which the universe may be divided into subtests and subtests assigned subsequently to examinees. The relative merits of alternative sampling plans for both items and examinees are discussed in Chapter 2.)

The results obtained from examinee-sampling and from multiple matrix sampling are used to estimate the parameters or characteristics of the N by K matrix—the results which would have been obtained by administering *all* items in the universe of interest to *all* examinees in the population. It must be remembered, however, that the N by K matrix illustrated in Figures 1-1 and 1-2 is in most cases a hypothetical matrix whose parameters are *estimated* from the subset of data gathered in practice through examinee-sampling or multiple matrix sampling.

Extreme Variations of Multiple Matrix Sampling Model

The three commonly used testing procedures labeled *item-sampling, examinee-sampling*, and *census testing* are extreme variations of the multiple matrix sampling model. In item-sampling, a sample of items (one subtest) is administered to the examinee population; in examinee-sampling, a sample of examinees responds to the item universe; and in census testing, the entire population responds to the entire item universe. Each of these three testing procedures is a variation of the matched-items model which is itself described more appropriately as a subclass of sampling plans within the multiple matrix sampling model. Of course, *item-examinee sampling*, in which one sample of items (one subtest) is administered to one sample of examinees (one subgroup), is an additional variation of the model.

Synonyms Encountered in the Literature

The terms *item-sampling, item-examinee sampling*, and *multiple matrix sampling* have often been used synonymously in the literature. To add to this confusion, multiple matrix sampling has been labeled by some authors as "gathering unmatched data" or "Type 1-2 sampling." Additionally, the procedure is referred to in some disciplines as a "partially balanced incomplete block design" or a "balanced incomplete block design" depending upon the method by which the item universe has been divided into subtests. Although there are several synonyms for *multiple matrix sampling*, none will be used here. They have been introduced merely to aid the reader who may choose to examine original literature sources cited within this book.

ADVANTAGES OF MULTIPLE MATRIX SAMPLING

Reduced Standard Error of Estimate

A concept important in discussing the advantages of multiple matrix sampling and one mentioned frequently both in this book and in the relevant literature is the *standard error of estimate* (SE). To explain what is meant by this term, consider the example where four experimental procedures have been developed for measuring weight and each procedure has been used to obtain in a standardized manner 1,000 independent measurements of the weight of a given object. Hypothetical measurements so acquired have been assembled into the frequency distributions given in Figure 1-3. The standard error of estimate associated with procedure *M* is the standard deviation (SD) of the 1,000 values for the weight obtained using procedure *M*; the standard errors of estimating the weight using procedures *N, O,* and *P* are determined identically. In Figure 1-3, the standard errors for procedures *N* and *O* are considerably larger than those for procedures *M* and *P*. These latter two procedures are said to be more *efficient* than the former. Additionally, there is another important difference. If the "true" weight of the object was 10 units, procedures *M* and *N* produce on the average estimates of the weight equal to the true weight. Such is not the case with procedures *O* and *P*. In this regard, the results obtained from procedures *M* and *N* are said to be *unbiased* estimates while results from the other two are *biased*. Specifically, procedure *O* produces estimates which are biased negatively while those from procedure *P* are biased positively. Procedure *M* is the best of the four because it produces estimates which are both unbiased *and* more efficient.

The differences among the standard errors of estimate depicted in Figure 1-3 illustrate an important advantage of multiple matrix sampling over examinee-sampling, item-sampling, and item-examinee sampling, namely, that the standard error of estimate per parameter will generally be less when the number of observations taken is held constant—defining the number of observations as the product of the number of examinees tested and the number of responses per examinee. For example, Lord and Novick (1968) have demonstrated algebraically that, when subtests are constructed by sampling items randomly and without replacement from the universe, the standard error in estimating the group *mean test score* using multiple matrix sampling is less than the standard error obtained with examinee-sampling. Furthermore, in estimating the mean test score, the minimum standard error of estimate under multiple matrix sampling is found by administering individually each item in the universe to a random sample of examinees—that is, by using 1-item subtests. Constructing 1-item subtests is the best procedure for estimating the mean universe score; however, it should be noted that this is not the best procedure for estimating the variance or any other central moment of the universe scores.

Although the standard error of estimate in multiple matrix sampling is less than that for item-sampling, examinee-sampling, or item-examinee samp-

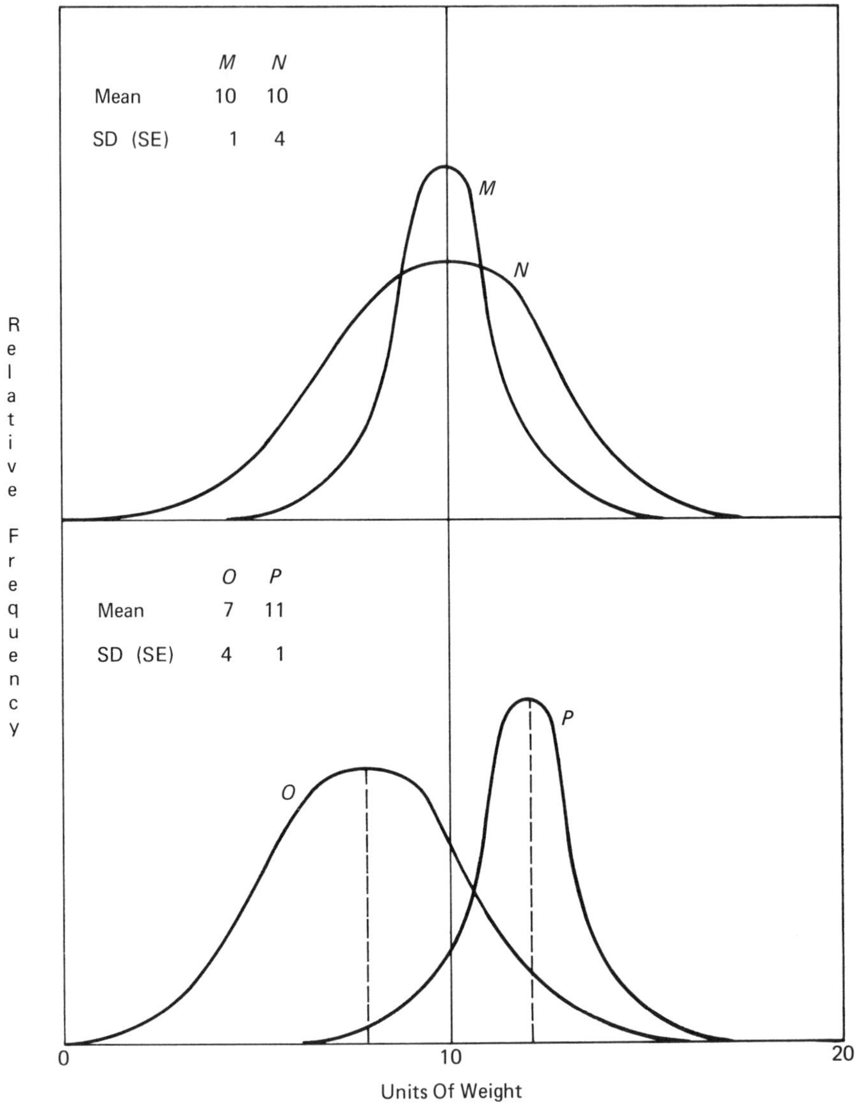

Figure 1-3. Hypothetical Distributions of Weight Measurements Resulting From 1000 Replications Using Four Measurement Procedures

ling, estimates of parameters obtained through multiple matrix sampling (and the other three procedures as well) are also unbiased. The relevant equations have been derived algebraically by Hooke (1956) and Lord (1960) and the results of several empirical investigations (e.g., Johnson and Lord, 1958; Lord, 1962; Plumlee, 1964; Cook and Stufflebeam, 1967; Shoemaker, 1970a, 1970b) using postmortem sampling support the unbiased properties of these estimators. In postmortem sampling an extant item-examinee matrix serves as the data base and the investigator estimates parameters of this base through multiple matrix sampling acting as if only certain items had been administered to certain examinees knowing all the while the results which were obtained by testing all examinees over all items. Although the equations given in Chapter 3 produce unbiased estimates of parameters, they should not be considered sacrosanct. It may be the case that for certain parameters (e.g., third and fourth central moments of the test score distribution whose estimators have a notoriously large standard errors of estimate) biased but more efficient estimators may be more useful, particularly if the amount and direction of bias were known.

Testing Large Item Universes

When the number of items in the universe is large, multiple matrix sampling provides the only feasible way by which the parameters of the universe scores may be estimated reasonably. Traditionally, the problem has been resolved by using either item-sampling or item-examinee sampling. The manner in which the parameters associated with large item universes are to be interpreted is discussed under the section entitled "Context Effect."

Reduction in Testing Time

For a given item universe, the testing time per examinee is reduced using multiple matrix sampling. This is an important consideration because it is frequently very costly or very difficult to obtain large blocks of testing time per examinee. Additionally, all examinees need not be tested. Because testing time is often synonymous with expense, the assessment of student performance over a given universe will cost less under multiple matrix sampling.

Additional Advantages

Several additional advantages of multiple matrix sampling are introduced here although they will be discussed in greater detail in subsequent chapters. An important point is that multiple matrix sampling as a procedure may be accepted more readily in certain situations than the matched-items design. In a school, for example, teachers fearing that test results may be used against them may be assured more convincingly if different students respond to different subtests. Secondly, given a limited amount of available testing time per examinee, performance on a larger item universe can be approximated through multiple matrix sampling than through the matched-items design. Thirdly, with multiple matrix sampling it is possible to estimate simultaneously parameters of several universes. Any given subtest may be a composite of several subtests. To the examinee, the subtest so constructed is merely another test; however, to the test constructor, the composite contains items from different universes each having parameters estimated through multiple matrix sampling. Fourthly, multiple matrix sampling as a measurement procedure is by no means applicable only to universes of achievement test items or to populations of examinees. As a procedure, it is applicable to a much broader spectrum of instruments or measurement techniques and to a wide range of populations. This point will be discussed in greater detail under the section entitled "Object-Variable Rationale." Lastly, and a point of some interest, it is possible that test administrators will not become as bored in individually administering different subtests as is sometimes the case when the same set of items is administered to all examinees.

LIMITATIONS OF MULTIPLE MATRIX SAMPLING

Context Effect

Multiple matrix sampling was proposed originally as a substitute procedure for estimating the normative distribution of scores on a standardized test. The amount of testing time required in norming a test had been traditionally substantial and it was thought that a procedure requiring less testing time per examinee would result in a more representative normative distribution because more individuals could be persuaded to participate. Although this argument is sound, it rests upon one critical assumption. Because each examinee responds to the entire test in practice, using multiple matrix sampling here assumes that any "context effect" is minimal, that is, that the response of examinees to a subtest are exactly those which would have been obtained had the examinees responded to the same items embedded in the complete test. Stated differently, the effect of the total test content and context on the response to an individual item should be maintained in each subtest. The validity of this assumption has received some attention in the literature (e.g., Owens and Stufflebeam, 1969; Shoemaker, 1970; Sirotnik, 1970; Huck and Bowers, 1972) and the results suggest that examinees do respond to items in subtests as they would respond to the same items in the complete test. The problem, however, has only recently been examined empirically and only a few tests have been used. For this reason, multiple matrix sampling should be used with caution here. This restraint will undoubtedly be eliminated in the future. Assuming that any context effect is primarily a function of the ratio of subtest length to total test length and the testing time per subtest, guidelines for specifying these two parameters could be determined empirically permitting the researcher to reduce any context effect to a predetermined and tolerable degree. It should be noted, however, that the context effect presents a problem only if multiple matrix sampling is being used to norm a test which, when used in practice, is given in its entirety. If performance on a standardized test was the metric by which the relative achievement of four examinee populations was to be determined, any context effect resulting from multiple matrix sampling would be constant across populations.

A related problem exists in estimating parameters of speeded tests or tests having a large speed component. A speeded test is one having items all of which are answered easily but whose time limit is very short. Such tests are designed primarily to measure rate of response and, without the restrictive time limit, it is assumed that every examinee could answer correctly each item. With a speeded test the investigator is interested primarily in the total correct score and the distribution of these scores for a particular group of examinees. Multiple matrix sampling may be used in this case to estimate the parameters associated with the test; however, deciding upon the appropriate time limit and number of items for an individual subtest is a difficult problem and one for which no guidelines exist currently in the literature.

There is another facet of the problem which requires careful consideration. Multiple matrix sampling has been proposed as a procedure for testing large item universes. Consider, for example, using the procedure to estimate the performance of kindergarten students with a 200-item universe. One plan for doing this would be to divide the 200-item universe into 10 subtests containing 20 items each and administering each subtest to 30 examinees. All examinees could handle easily a subtest of 20 items, yet few, if any, could respond to all 200 items when administered as one test. However, the parameters estimated would be those associated with the 200-item universe. In this situation, two questions come immediately to mind. First, how are the estimates of these parameters interpreted? Second, is the context effect problem here similar to that encountered when estimating the normative distribution associated with a standardized test? In reply to the first question, the parameters are those associated with the 200-item universe when the universe is administered to examinees under ideal testing conditions. In an ideal testing situation, an examinee's response to an item is a function of his ability level and not a function of factors associated with the testing environment such as item sequence, test length, and the time limit of the test. Because the effect of factors such as these on item performance is *reduced* by using subtests instead of the entire universe, a more accurate assessment of examinee performance is obtained. The answer to the second question, therefore, is that the problem is just the opposite. When measuring student achievement on an item universe, it is desirable to reduce the effect of the universe of items on the performance of a single item; when using multiple matrix sampling to estimate the parameters associated with a standardized test, performance on an item must reflect the context and influence of the total test.

Sampling of Examinees

The equations used to estimate parameters in multiple matrix sampling assume that each subtest is administered to a subgroup of examinees each of whom has been selected randomly from the population. To do this, of course, requires an exhaustive listing of the entire population of interest. Because this is frequently a difficult task, other methods of selecting examinees (e.g., two-stage sampling and cluster sampling) are often used. In two-stage sampling, for example, a specified number of examinee subgroups are sampled randomly from the population of subgroups in the first stage. Then, in the second stage, examinees are sampled randomly from within each subgroup sampled originally. Examples of populations of subgroups used frequently are classrooms, schools, and school districts. The point here is that these examinee-sampling plans violate one assumption upon which estimators currently available in multiple matrix sampling have been developed, and using them introduces a degree of bias into the results. Although the amount of such bias will decrease as the number of examinees per subtest is increased, this problem, nevertheless, reflects a current limitation in the methodology of multiple matrix sampling. When such examinee-sampling procedures are incorporated into the derivation of estimators, it can be anticipated that using them will permit fewer examinees being tested to estimate a given parameter with a prescribed degree of accuracy.

Estimating the Universe Score for Each Examinee

Multiple matrix sampling to date has been concerned primarily with estimating the parameters associated with a distribution of universe scores. Very little attention has been given to using the procedure to estimate the universe score for each individual examinee. Because so few of the necessary equations are available currently, any researcher concerned primarily with obtaining universe scores for each examinee should use some variation of the matched-items model. It should be noted, however, that Kleinke (1972, 1973) has attempted to expand the use of multiple matrix sampling into the assessment of individual differences by using a linear regression model to predict the universe score for each examinee from his subtest score. Although his interest was in approximating the distribution of universe scores and not in the assessment of individual differences, his preliminary results suggest the possibility that multiple matrix sampling may be used upon refinement in the latter case. An evaluation of his "*c*-prediction model" is perhaps best made in the light of subsequent research. Of particular importance here, of course, is the standard error of estimate associated with each predicted universe score. Das (1967) has also described a procedure for using multiple matrix sampling to classify individual students into ability categories. In his procedure, each student is administered at least three subtests having items selected randomly from the universe. Although little has been done to date, the reader can anticipate multiple matrix sampling being adapted in the future to individual assessment.

Logistics of Multiple Subtests

A potentially serious limitation of multiple matrix sampling is in the logistics of giving different tests to different subgroups of examinees. If the instructions to each item are written on the test booklet, administering different tests to different examinees within the testing room is accomplished with relative ease. If test items are administered individually, the problem is also minimal. If, however, the test is group-administered and each item requires oral instructions by the administrator, the arrangement is more complex. In this situation, the examinees must be arranged and isolated according to subtest.

Necessary Statistical Procedures

Although all computational procedures used in multiple matrix sampling are detailed in Chapter 3, it should be noted that the analysis of subtest results in estimating parameters of the universe for a given population will be extremely laborious without the use of an electronic computer. All computations in multiple matrix sampling have been computerized by the author and the necessary program and documentation are given in Appendix A. The reader using multiple matrix sampling should anticipate using this program and should use a desk calculator if and only if no access to a computer exists.

MAJOR AREAS OF APPLICATION

Multiple matrix sampling is suited ideally to a variety of measurement problems. The range of applications has been obscured by the fact that most of the research on the topic has been cast within the framework of achievement testing in education. This is potentially misleading because the procedure is applicable to a much broader spectrum of measurement problems. Indeed, the number of applications is restricted only by the degree of ingenuity on the part of the individual researcher.

Object-Variable Rationale

To demonstrate the wide applicability of multiple matrix sampling, consider the matrix which serves as the data base for the procedure. Up to this point the matrix has always been an examinee (rows) by item (columns) matrix. A more appropriate way to consider the data base is an *object* (rows) by variable (columns) *matrix*. When viewed from this perspective, examinee by item is seen to be but a special case of object by variable. There is no reason why resident (row) by survey question (column), automobile (row) by smog control test (column), fish (row) by disease symptom (column), and rock (row) by hardness characteristic (column) cannot be valid matrices for this procedure. Because numerous measurement problems can be considered within the framework of an object by variable matrix, potential applications of multiple matrix sampling extend far beyond educational research.

Examples of Specific Applications

Although multiple matrix sampling has potential applications in numerous fields, those relating to education and educational research have been the most prominent and the most widely discussed to date. The following are several areas where the procedure has immediate applicability in education:

Curriculum Evaluation. At the local, state, and national level, multiple matrix sampling may be used in conjunction with a relevant item universe to evaluate any instructional program. In addition to the increased content coverage afforded over standardized tests, the procedure reduces both the testing time per examinee and the number of examinees required for testing.

Evaluation of Competing Instructional Programs. Given an item universe which spans comprehensively the major content categories of a particular field of instruction, multiple matrix sampling may be used to assess student achievement over this universe and content substrata within it for each instructional program. The resulting parameter estimates provide the basis for determining the relative strengths and weaknesses of each program. The specific procedure by which multiple matrix sampling is used in this case and in curriculum evaluation generally is discussed in greater detail in Chapter 6.

Hypothesis Testing. Parameters estimated through multiple matrix sampling are integrated easily into a variety of statistical hypothesis testing procedures. For example, the item scores obtained from each subtest are amenable to completely randomized and factorial analysis of variance designs. This topic is the subject of Chapter 5.

Questionnaires and Surveys. Two obvious extensions of multiple matrix sampling are estimating parameters associated with questionnaires and surveys. Preliminary research in this area is promising and is discussed in Chapter 6.

Attitude Measurement. Attitudes may be scaled through multiple matrix sampling and the procedure has been used successfully with items in paired-comparison format and with Likert items.

Standardized Tests. Multiple matrix sampling may be used to estimate the normative distribution for a standardized test. The very length and time required by a standardized test often make it difficult if not impossible for the test developer to norm the instrument on a truly representative sample from the population of interest. The time required per student is frequently difficult to obtain and more cooperation may result through multiple matrix sampling because it requires less testing time per student. It should be noted that the multiple matrix sampling design was proposed originally by William Turnbull and Robert Ebel as one most appropriate for estimating the mean test score of the normative distribution for a standardized test. There is one problem, however, with this application of the procedure which should be mentioned. Most standardized tests are slightly speeded and the manner in which this feature should be incorporated into the timing of individual subtests is unresolved currently. The problem is not insurmountable and appears to be one readily resolved empirically.

Information about individual test items is acquired easily through multiple matrix sampling and situations where this is important are numerous. In testing for college admission, for example, forms of a new test must be kept secure until the date of their use—yet statistical information on individual items is needed to assemble various forms of the test. Similarly, the National Foundation for Educational Research in England and Wales has used multiple matrix sampling to pretest universes of items which are then stored in "item banks." This use of the procedure has been described by Wood and Skurnik (1969).

These six areas span comprehensively the uses to which multiple matrix sampling has been put in recent years. It is anticipated that this list will lengthen in the future. The rate of growth will undoubtedly be a reflection of the degree to which researchers become aware that frequently—more frequently than is often realized—the prime thrust of an investigation is assessing *group* performance, not individual performance, and that for this case the matched-items model is ill-suited. There is an alternative procedure called multiple matrix sampling which is for many measurement problems (but not all) the more appropriate procedure.

Chapter Two

Procedural Guidelines

Multiple matrix sampling as a procedure involves basically three steps: (a) subdividing an item universe into subtests, (b) administering each subtest to a group of examinees selected from the population, and (c) estimating parameters of the item universe from subtest results. Although the general procedure is described easily, implementing it, that is, performing the first two steps, produces many interesting questions. For example: How many subtests should be constructed? To how many examinees should each subtest be administered? Is it more appropriate to administer a few subtests each containing a large number of items or a large number of subtests each containing a few items? These are only a few of the questions encountered frequently when using multiple matrix sampling and their diversity suggests correctly that there are numerous ways in which the procedure may be used in a particular project. Described in this chapter are a set of general guidelines and concomitant discussion which hopefully will aid the researcher in selecting sampling plans for both items and examinees. Although the guidelines cited here do represent the accumulated experiences of several investigators in using multiple matrix sampling, they are by no means complete and exhaustive. As more experience is gained with the procedure, it is assuredly the case that additional guidelines will be generated. It is in this area that one of the major research thrusts in multiple matrix sampling can be expected to occur in the future. The third step mentioned above is the content and substance of Chapter 3.

GENERAL CONSIDERATIONS IN SAMPLING

A distinction of importance in sampling items from the universe and examinees from the population is the difference between sampling *with replacement* and sampling *without replacement*. Consider the example in which two subtests containing 5 items each are to be constructed from a 10-item universe. If the items for each subtest were selected randomly but with replacement, the 5 items selected for the first subtest would be returned to the universe *before* the sampling of the 5 items for the second subtest. In this case, it is possible for a given item or set of items in the universe not to be included in either subtest or for one or more items to be included in both subtests. If the items for each subtest had been sampled *without replacement*, the 5 items selected for the first subtest would not be returned to the universe before selecting items for the second subtest and, in this example, the items for the second subtest would be the remaining 5 items in the universe. As is the case with items and is also true with examinees, the terms *with replacement* and *without replacement* always refer to among subtests or among examinee subgroups and do not refer to individual items or individual examinees. This means simply that items within a subtest or examinees within a subgroup are always sampled randomly and without replacement. Were this not the case, the same item could be found more than once in the same subtest or the same examinee included more than once in the same subgroup.

A particular sampling plan in multiple matrix sampling may be denoted by ($t/k/n$) where t refers to the number of subtests, k to the number of items per subtest, and n, to the number of examinees to which each subtest is administered. The number of items in the universe is denoted here as K and the number of examinees in the population as N. In multiple matrix sampling it is not necessary that the number of items per subtest be the same for each subtest or that each subtest be administered to the same number of examinees. Although the computational formulas detailed in Chapter 3 do not require equal k's and n's across subtests, the sampling plan should be constructed initially so that these equalities prevail. In multiple matrix sampling, the results from individual subtests are combined to obtain the single best estimate of each parameter. When the k's and n's are equal (or approximately equal) across subtests, the results of each subtest contribute equally in obtaining the pooled estimate of the parameter; when unequal, the weight of each subtest in the composite is, for subtest s, equal to the product $k_s n_s$. In the majority of investigations using multiple matrix sampling there is initially no reason for differentially weighting subtest results and the problem arises typically only after the data have been gathered and one finds that subtests were not assigned properly to examinees. Regarding the number of subtests, there is no computational requirement in multiple matrix sampling that $t = K/k$, that is, that all items in the universe be assigned to some subtest. Although unnecessary computationally, in most projects all items are tested because the investigator has constructed his item universe carefully and is interested in gathering the maximum amount of information about his universe. Indeed, on occasion it is desirable to construct subtests such that the product tk is greater than K, resulting in individual items being included in more than one subtest. If the item universe were small, the investigator might increase the number of items per subtest to take full advantage of the testing time available for each student. When this is done, tk should be an integer multiple of K and items should be sampled randomly but subject to the restriction that, across subtests, each item appears with equal frequency. For example, if a (10/20/30) sampling plan were used with a 50-item universe, the multiple is 4 and each of the 50 items should appear in exactly 4 subtests.

Although it is most frequently the case that all the items in the universe are tested, it is infrequently the case that every member of the population is tested. Once again, there is no computational necessity for testing every member in the population of interest.

Sampling Of Items. The items for each subtest may be sampled from the universe either through a *random sampling plan* or a *stratified-random sampling plan*. Using the latter, the items in the universe are subdivided initially into categories (or strata) and, in selecting items for each subtest, a given number of items are sampled randomly from each stratum. One criterion used commonly for determining the number of items to select from each stratum for each subtest is the ratio (proportion) of stratum to universe. For example, if a 100-item universe were divided into 4 strata consisting, respectively, of 60, 20, 10 and 10 items, .6 of the items for each subtest would be sampled randomly from the first stratum, .2 from the second, and .1 from each of the remaining two. Here, each subtest reflects the general characteristics of the universe as defined by the method of stratification. In multiple matrix sampling, a stratified-random sampling plan is preferred over a random sampling plan because the generalizability of scores obtained on subtests constructed through the former procedure is greater than the latter. Generalizability here refers to the relationship between scores on one subtest and any other subtest for a given examinee. The theoretical development and concomitant empirical research supporting this conclusion has been given by Rajaratnam, Cronbach, and Gleser (1965) and Shoemaker and Osburn (1968).

It is frequently the case with large item universes that the estimated parameters for the total universe are difficult to interpret. Consider, for example, a 600-item universe whose items are designed to measure student performance in a bilingual Spanish/English language program. Assume that half of the 600 items are in English with the remaining in Spanish. In place of estimating parameters of the 600-item universe, the investigator will obtain more useful information by considering the English items and the Spanish items as two separate universes with individual parameters estimated through multiple matrix sampling. The point here is that a given universe may be—and frequently should be—thought of as a composite of several subuniverses. When this is done, the entire multiple matrix sampling procedure is applied to each subuniverse of interest. If the researcher plans initially to estimate the parameters of several subuniverses, the items must be allocated accordingly through a stratified sampling plan with the additional stratification being by subuniverse. For example, if the researcher were interested in estimating separately parameters of the 300-item universe for English items and the 300-item universe for Spanish items, half of the items in each subtest would be in each language (i.e., items stratified according to subuniverse). Additionally, the items for each language would be selected according to a stratified-random sampling plan within each language. In the computer program given in Appendix A for estimating parameters through multiple matrix sampling, one option available to the user is storing on magnetic tape or disk all item scores obtained by each examinee on each subtest. The user may then make repeated "passes" through this data base with each pass selecting scores on those items which form a particular subuniverse of interest. All details for doing this are given in Appendix A.

There is an additional complication in the sampling of items from the universe of which the researcher must be aware. With some universes, subsets of items constitute an item block and the researcher must decide whether to treat the items as a whole during the sampling process or to deal individually with each item. This problem arises, for example, in measuring reading comprehension when a subset of questions is associated with one particular paragraph or story. From a practical point of view, most researchers would be inclined to retain the items as a block—particularly if the paragraph or story were long. However, from a statistical viewpoint, it is well known that any item sampling procedure which is not random will produce a biased estimate of each parameter of interest. The amount of bias is undoubtedly a function of the number of items within each block and the number of blocks. This problem to date has generally been ignored but, nevertheless, is deserving of some attention. Until the problem is resolved either empirically or algebraically, it is probably safe to maintain the block of items when the number of items within the block is small and few blocks are present within the universe of interest.

Sampling Of Examinees. Examinee populations like item universes may also be stratified and, when considered in this manner, provide a more reasonable basis for forming examinee subgroups having a high degree of homogeneity among themselves with each representative of the population. Any procedure which produces heterogeneous examinee subgroups should be avoided because of the decrease in generalizability which it introduces into the results. The results of a particular subtest should be indicative of the population and not a particular segment of the population. The point here is that examinees should be assigned to subtests using a stratified-random sampling plan when stratification is possible.

It may be that the investigator is interested in estimating parameters for a universe (or subuniverse) individually for subpopulations within the population. For example, two subpopulations which might be of interest are males and females. Were this the case, assigning examinees to subtests would involve an additional stratification, namely, according to subpopulation. The general procedure here is identical to that used to estimate parameters for subuniverses.

GENERAL CONSIDERATIONS IN ESTIMATING PARAMETERS

Minimum Number of Items Per Subtest. The choice of a sampling plan most appropriate for a particular project is a function of the parameters being estimated. From a computational viewpoint, a minimum of r items is required per subtest to estimate the rth moment of the distribution of universe scores. For example, if the researcher were interested in estimating the fourth central moment, each subtest must contain at least four items; for the second central moment (variance), each subtest must contain minimally two items. This is a mathematical requirement related directly to the equations used to estimate moments given in Chapter 3. Lord and Novick (1968) have demonstrated algebraically and several investigators (e.g., Shoemaker, 1972; Barcikowski, 1972) have verified empirically that, keeping the number of observations acquired constant, the most efficient estimate of the universe mean test score is obtained by using a $(K/1/n)$ sampling plan, that is, by constructing K subtests each containing 1 item. The point here is this: although a $(K/1/n)$ plan is optimal for estimating the mean test score, it cannot be used to estimate any other moment. This is an important point because it is infrequently the case that a researcher is interested solely in estimating the mean performance over an item universe. In each investigation, the researcher must decide initially the highest moment of interest to him and then select a sampling plan which estimates that moment with reasonable accuracy. For example, if a researcher were interested in approximating the entire normative distribution of test scores with the negative hypergeometric distribution, he must select a sampling plan which estimates accurately the variance of the test score distribution. (As will be explained in Chapter 3, the negative hypergeometric distribution is a function of the number of items in the universe, the estimated mean universe score, and the estimated variance of universe scores.) It should be noted that this rule of thumb is offered in the absence of any computational procedure for determining the number of items required to estimate a given moment (beyond the mean and variance) with a prescribed degree of accuracy.

Estimating Parameters With Prescribed Accuracy

One procedure for determining the number of items per subtest required to estimate a given parameter with a specified degree of accuracy is that of taking the equation for estimating the standard error associated with that parameter, inserting the acceptable standard error, and solving for the number of items per subtest. As an example of what is meant here, take Equation (3.15) for the standard error of the pooled estimate of the mean universe score given in Chapter 3. Assuming that $t = K/k = N/n$ and solving for t results in

$$t = 1 + \frac{(N - 1)SE(\hat{\mu}'_{1p})^2}{\mu_2(1 - KR_{20})} \tag{2.1}$$

where KR_{20} refers to the Kuder-Richardson Formula 20 reliability coefficient. Equation (2.1) assumes, of course, that prior estimates of certain parameters (i.e., μ_2 and KR_{20}) are available to the researcher and also that he has some acceptable standard error of estimate for the mean universe score in mind. The problem may possibly be reduced in complexity by using approximations of parameters. For example, Lord (1959) has determined empirically that $\mu_2(1 - KR_{20})$ is equal approximately to .187 K for achievement tests having among themselves a wide variation in parameters. (Gardner [1970] has demonstrated theoretically why this approximation is reasonable and McMorris [1972] has verified empirically its accuracy and the feasibility of additional "short-cut" estimates of parameters.)

This general procedure for estimating t and subsequently k appears promising. Implementing it, however, for a particular parameter of interest requires that the equation for the standard error of the pooled estimate of that parameter has been derived. Herein lies the problem because deriving such equations is not a casual undertaking. Although additional equations for standard errors of pooled estimates will be derived undoubtedly in the future, only Equations (3.15) and (3.17) in this book have been reported to date. The first is for the pooled estimate of the mean universe score; the second, for the pooled estimate of the variance of universe scores. This line of research is important because it will provide an algebraic solution for determining at the beginning of an investigation the sampling plan ($t/k/n$) required to estimate a given parameter with a predetermined degree of accuracy. Until such equations are derived, guidelines for selecting an appropriate sampling plan will have to be determined empirically. The two techniques will complement one another, because that which is determined empirically should support and, indeed, provide a test for that which is determined algebraically.

Beforehand Approximations of Parameters. An important point in selecting a sampling in multiple matrix sampling is that the choice is made easier if approximations of various parameters are known in advance. This was the case with (2.1) and is implied in the guidelines given in this chapter. Although the reasoning may appear initially circuitous, what is meant here by "approximation" is "a rough approximation." The researcher is not required to estimate, for example, the index of skewness but only whether the normative distribution of test scores will be symmetrical or skewed markedly. Approximations of this type can be made by most researchers. If the performance of a group of students on a standardized achievement test were being estimated through multiple matrix sampling, the available norms for the test may be used to roughly approximate parameters. For other item universes, the researcher might obtain rough estimates of parameters by pilot-testing certain segments of the item universe. Also, as experience is gained with a particular item universe over time, rough approximations of parameters can be made easily. An alternative procedure for selecting a sampling plan would be to simulate the entire process using estimates of parameters and then pick the sampling plan which has the smallest standard error for the highest moment of interest. (In Chapter 4, a simulation model and accompanying computer program which may be used for this purpose are described in detail.) Initial estimates of the characteristics of the item universe are important in selecting a sampling plan and the researcher should take advantage of them whenever possible.

GENERAL GUIDELINES

The guidelines given here are the result generally of postmortem item-examinee sampling investigations involving data bases whose parameters have been manipulated within a computer simulation model. The criterion used in contrasting the effectiveness of individual sampling plans in these investigations has been the standard error of estimate for each parameter. Examples of investigations which illustrate well this general procedure are those reported by Barcikowski (1972), Shoemaker (1972), and Shoemaker and Osburn (1968). Additional guidelines have been derived algebraically by Osburn (1969) and by Lord and Novick (1968). The guidelines given below refer to the relative magnitudes of standard errors of estimate. For example, the seventh guideline resulted from the observation that, for a given sampling plan, the standard error of estimate for each parameter will generally be less if the normative distribution of universe scores is markedly skewed than if the scores are distributed normally. This is true for variations in size of the item universe and for a variety of sampling plans. Stating this finding in the obverse results in the guideline. No mention, however, is made of the magnitude of any specific standard error of estimate. In multiple matrix sampling, several sampling plans are usually feasible in any project and the selection of a particular sampling plan is the result typically of both practical and statistical considerations. The following are general guidelines resulting from several investigations in which the problem has been attacked both empirically and algebraically:

1. The number of observations (defined as the product tkn) acquired by the sampling plan is an important variable. In general, as the number of observations increases, the standard error of estimating parameters decreases.

(The major exception to this guideline occurs when guideline 4 is not followed.)

2. Items and examinees should be sampled without replacement from the universe and population.
3. If possible, the product tk should always be equal to or greater than K, that is, all items in the universe of interest should be included in some subtest.
4. When the product tk is greater than K, tk should be an integer multiple of K and items should be selected randomly but subject to the restriction that among subtests each item appears with equal frequency.
5. Increasing the number of examinees per subgroup is least effective in reducing the standard error of estimate.
6. For normal normative distributions, increases in the number of items per subtest are most effective in reducing standard errors of estimate; for negatively-skewed distributions or positively-skewed distributions, increases in the number of subtests are most effective.
7. In general, fewer observations are required to estimate parameters of a skewed normative distribution than of a normal normative distribution.
8. If subtest items are being selected according to a stratified-random sampling plan instead of a random sampling plan, items should be stratified according to difficulty level and not according to content.
9. As the reliability of the normative distribution of test scores increases, it becomes increasingly difficult to estimate parameters. For this reason, it is true generally that a relatively large number of observations is required by the sampling plan when estimating parameters of a distribution having high reliability. This is true also when the variance of item difficulty indices is large.
10. If confidence intervals for pooled estimates of parameters are to be computed using jackknifed standard errors of estimate (a procedure described in Chapter 3), the number of subtests should be at least ten.

Guidelines 1 through 10 assume a certain amount of statistical familiarity with the universe whose parameters are being estimated. In those situations where the investigator is uncertain about the general limits of the parameters associated with the universe of interest, the following rules-of-thumb will generally suffice in selecting a particular sampling plan ($t/k/n$): (a) Always have at least 6 to 8 subtests, (b) Test all items in the universe, and (c) Have more items per subtest and fewer subtests than many subtests with few items each.

MULTIPLE MATRIX SAMPLING STEP-BY-STEP

Step 1: Construct or select the K-item universe. If possible, assemble the items into strata according to difficulty level.

Step 2: Determine the limitations and restrictions which must be imposed upon the test administration procedure.

Step 3: Select a sampling plan ($t/k/n$) which is appropriate in view of the estimated characteristics of the normative distribution, the restrictions and limitations inherent in the test administration procedure, the desired degree of accuracy in the results, and guidelines 1 through 10.

Step 4: Administer subtests to examinees in a standardized procedure. Avoid confounding subtest with examinee subgroup, that is, make every attempt to have homogeneous examinee subgroups.

Step 5: Compute estimates of parameters using the computer program given in Appendix A.

Chapter Three

Computational Formulas

Computational formulas used in multiple matrix sampling are described in this chapter and their use is demonstrated in a project designed to assess the spelling proficiency of kindergarten students. The listing of formulas here is not exhaustive as undoubtedly additional formulas will be derived and used in the future. It should be noted initially that all formulas cited within this chapter assume only that the universe score for each examinee is an unweighted and linear sum of his item scores.

ASSESSMENT OF SPELLING PROFICIENCY

The computations involved in multiple matrix sampling are perhaps best enumerated through applying the procedure to a specific example. The following example reported originally by Shoemaker and Okada (1970) has been selected for this purpose and concerns the assessment of spelling proficiency in kindergarten students. Although the universe in this example consisted of 50 items, it is generally the case that larger item universes will be used in most investigations employing multiple matrix sampling.

A spelling program was being designed for kindergarten students and the word and rule content of this program was to be related closely to the reading program used by these students. Before constructing such a program it was necessary to determine the spelling proficiency of those students who have used the reading program but have not had formal spelling instruction on the related words. Although there were 78 unique words introduced in the particular reading program under consideration, technical considerations dictated that only words having regular spellings be included in the item universe. As a result, the original universe was reduced from 78 words to 50 words. The modified universe was then subdivided through random sampling without replacement into five subtests containing ten words each. (This is one of many procedures which could have been used. Alternative procedures are discussed in detail in Chapter 2.)

Three kindergarten classes were selected randomly from the pool of nine available classes. Students within each class were divided at random into five groups and each group was assigned at random to one of the five subtests. Each test was administered individually and all items were scored dichotomously (1 = pass, 0 = fail).

USE OF SUBTESTS IN MULTIPLE MATRIX SAMPLING

In multiple matrix sampling, subtests results are of secondary interest. Of chief concern is the estimation of parameters and the standard error of estimate associated with each parameter. (It is very difficult to interpret meaningfully estimated parameters without having some idea about the amount of variability associated with the estimate.) In this example, the parameters of interest are the characteristics of the universe scores which would have been obtained had all students in all nine kindergarten classes been tested over the entire set of 50 spelling items defining the universe of interest.

Subtest results are important in that the results from *each* subtest may be used to provide an estimate of each parameter of interest. For example, from subtest 1 (containing ten spelling items) it is possible to obtain an estimate of the mean universe score on the 50-item universe, the variance of these scores, the third and fourth central moments of universe scores, the index of skewness, the degree of kurtosis, and the reliability of the universe scores. All of these parameters, of course, are not independent but each can be estimated from the results of *one* subtest. (If there are fewer than four items in the subtest, only certain of these parameters may be estimated. This problem is described in Chapter 2.) For each parameter of interest, the general procedure followed in multiple matrix sampling is to (a) estimate the parameter from the results of each subtest, and then (b) combine these independent estimates of the parameter into a single "best" estimate. This "pooled" estimate of each parameter is preferred over an estimate obtained from a subtest because, on the average, it will approximate more closely the true value of the parameter, i.e., its standard error of estimate will be less. In this example, five subtests were used and, hence, five estimates were made of each parameter.

ESTIMATING PARAMETERS FROM SUBTEST RESULTS

All of the estimation procedures described in this chapter are computerized easily and the program developed by the author which performs all of the necessary computations is given in Appendix A. Because the computations involved in multiple matrix sampling are both numerous and tedious, the reader should plan on using the computer program and *not* a desk calculator.

Estimation of Moments

In estimating the moments of the universe scores from the item scores obtained on one subtest, the following four equations derived by Pandey and Shoemaker (1973) are used. Each of these equations was derived within the mathematical framework established originally by Hooke (1954, 1956a, 1956b). Casting the item scores obtained by each examinee into an examinee (rows) by item (columns) matrix and denoting this matrix by $\| X_{ij} \|$, the estimate of the mean universe score obtained from subtest s is

$$\hat{\mu}'_{1_s} = \frac{K \sum_{i=1}^{n} \sum_{j=1}^{k} X_{ij}}{nk} \tag{3.1}$$

Estimating the variance, third central moment and fourth central moment of universe scores requires a series of preliminary calculations. Two additional matrices Y and Z are required where $Y = \| Y_{ij} \| = \| X_{ij}^2 \|$ and $Z = \| Z_{ij} \| = \| X_{ij}^3 \|$. The general procedure used to obtain estimates of these central moments from the results of one subtest is similar and the procedure is demonstrated most easily for the variance. The preliminary calculations involved in estimating the variance are indicated in Tables 3-1, 3-2, and 3-3. The Sigmas (Σ 's) defined in Table 3-1 are computed initially where the plus sign (+) given as a subscript indicates that the subscript replaced by the + is summed over all values; secondly, the d-statistics defined in Table 3-2; and, thirdly, the D-statistics defined in Table 3-3. Within this framework,

$$\hat{\mu}_{2_s} = \frac{(N-1)}{N} \left\{ D_2 + \frac{D_4}{K} \right\} K^2 \tag{3.2}$$

Table 3-1. Sigmas (Σ's) Associated With Table 3-2

Σ	*Arithmetic Expression*
1	$(X_{++})^2$
2	$\sum_{i=1}^{n} X_{i+}^2$
3	$\sum_{j=1}^{k} X_{+j}^2$
4	$\sum_{i=1}^{n} \sum_{j=1}^{k} X_{ij}^2$

Table 3-2. Conversion Of Sigmas (Σ's) To *d*-Statistics

		Σ			
d	*Constant*	*1*	*2*	*3*	*4*
1	$1/(n^{[2]}k^{[2]})$	1	−1	−1	1
2	$1/(n\quad k^{[2]})$		1	0	−1
3	$1/(n^{[2]}k\quad)$			1	−1
4	$1/(n\quad k\quad)$				1

Table 3-3. Computation of *D*-Statistics From *d*-Statistics

	d			
D	*1*	*2*	*3*	*4*
1	1			
2	−1	1		
3	−1	0	1	
4	1	−1	−1	1

To demonstrate some of the computations involved in Tables 3-2 and 3-3,

$$d_1 = \left[(1)(\Sigma_1) + (-1)(\Sigma_2) + (-1)(\Sigma_3) + (1)(\Sigma_4)\right]\left[\frac{1}{n^{[2]}k^{[2]}}\right]$$

where

$$n^{[i]} = n(n-1)(n-2)\ldots(n-i+1).$$

Also,

$$D_3 = (-1)(d_1) + (0)(d_2) + (1)(d_3).$$

Tables 3-4, 3-5, and 3-6 indicate the preliminary computations involved in estimating the third central moment of universe scores. Using these three tables,

$$\hat{\mu}_{3_S} = \frac{(N-1)(N-2)}{N^2} \left\{ A_4 + \frac{3A_8}{K} + \frac{A_{10}}{K^2} \right\} K^3 \qquad (3.3)$$

Table 3-4. Sigmas (Σ's) Associated With Table 3-5

Σ	*Arithmetic Expression*
1	$(X_{++})^3$
2	$(\sum_{i=1}^{n} X_{i+}^2)(X_{++})$
3	$(\sum_{j=1}^{k} X_{+j}^2)(X_{++})$
4	$\sum_{i=1}^{n} X_{i+}^3$
5	$\sum_{j=1}^{k} X_{+j}^3$
6	$\sum_{i=1}^{n} \sum_{j=1}^{k} X_{ij}X_{i+}X_{+j}$
7	$(Y_{++})(X_{++})$
8	$\sum_{i=1}^{n} Y_{i+}X_{i+}$
9	$\sum_{j=1}^{k} Y_{+j}X_{+j}$
10	$\sum_{i=1}^{n} \sum_{j=1}^{k} X_{ij}^3$

The fourth central moment requires Tables 3-7, 3-8, and 3-9 and the estimate is given by

$$\hat{\mu}_{4_S} = \left[\frac{3(N-1)^2}{N^2} \left\{ F_4 + \frac{4F_{13}}{K} + \frac{2F_{17}}{K^2} + \frac{2F_{18}}{K} + \frac{4F_{25}}{K^2} + \frac{F_{27}}{K^2} + \frac{F_{30}}{K^3} \right\} + \frac{(N-1)(N^2-3N-3)}{N^3} \left\{ F_8 + \frac{2F_{22}}{K} + \frac{3F_{29}}{K^2} + \frac{4F_{31}}{K^2} + \frac{F_{33}}{K^3} \right\} \right] K^4 \qquad (3.4)$$

Table 3-5. Conversion of Sigmas (Σ's) to a-Statistics

		Σ									
a	*Constant*	*1*	*2*	*3*	*4*	*5*	*6*	*7*	*8*	*9*	*10*
1	$1/(n^{[3]}k^{[3]})$	1	−3	−3	2	2	6	3	−6	−6	4
2	$1/(n^{[2]}k^{[3]})$		1	0	−1	0	−2	−1	3	2	−2
3	$1/(n^{[3]}k^{[2]})$			1	0	−1	−2	−1	2	3	−2
4	$1/(n\ k^{[3]})$				1	0	0	0	−3	0	2
5	$1/(n^{[3]}k\ )$					1	0	0	0	−3	2
6	$1/(n^{[2]}k^{[2]})$						1	0	−1	−1	1
7	$1/(n^{[2]}k^{[2]})$							1	−1	−1	1
8	$1/(n\ k^{[2]})$								1	0	1
9	$1/(n^{[2]}k\ )$									1	−1
10	$1/(n\ k\ )$										1

Table 3-6. Computation of A-Statistics from a-Statistics

	a									
A	*1*	*2*	*3*	*4*	*5*	*6*	*7*	*8*	*9*	*10*
1	1									
2	−1	1								
3	−1	0	1							
4	2	−3	0	1						
5	2	0	−3	0	1					
6	1	−1	−1	0	0	1				
7	1	−1	−1	0	0	0	1			
8	−2	3	2	−1	0	−2	−1	1		
9	−2	2	3	0	−1	−2	−1	0	1	
10	4	−6	−6	2	2	6	3	−3	−3	1

Table 3-7. Sigmas (Σ's) Associated With Table 3-8

Σ	*Arithmetic Expression*	Σ	*Arithmetic Expression*
1	$(X_{++})^4$	17	$\sum_{j=1}^{k}(\sum_{i=1}^{n} X_{ij})^2 + \sum_{j=1}^{k}\sum_{\substack{h=1\\ j\neq h}}^{k}(\sum_{i=1}^{n} X_{ij}X_{ih})^2$
2	$(X_{++})^2 \sum_{i=1}^{n} X^2_{i+}$	18	$(Y_{++}) \sum_{i=1}^{n} X^2_{i+}$
3	$(X_{++})^2 \sum_{j=1}^{k} X^2_{+j}$	19	$(Y_{++}) \sum_{j=1}^{k} X^2_{+j}$
4	$(\sum_{i=1}^{n} X^2_{i+})^2$	20	$(X_{++}) \sum_{i=1}^{n} Y_{i+}X_{i+}$
5	$(\sum_{j=1}^{k} X^2_{+j})^2$	21	$(X_{++}) \sum_{j=1}^{k} Y_{+j}X_{+j}$
6	$(X_{++})(\sum_{i=1}^{n} X^3_{i+})$	22	$\sum_{i=1}^{n} X^2_{i+}Y_{i+}$
7	$(X_{++})(\sum_{j=1}^{k} X^3_{+j})$	23	$\sum_{j=1}^{k} X^2_{+j}Y_{+j}$
8	$\sum_{i=1}^{n} X^4_{i+}$	24	$\sum_{i=1}^{n} Y_{i+}(\sum_{j=1}^{k} X_{ij}X_{+j})$
9	$\sum_{j=1}^{k} X^4_{+j}$	25	$\sum_{j=1}^{k} Y_{+j}(\sum_{i=1}^{n} X_{ij}X_{i+})$
10	$(X_{++}) \sum_{j=1}^{k}(\sum_{i=1}^{n} X_{ij}X_{i+})X_{+j}$	26	$\sum_{i=1}^{n}\sum_{j=1}^{k} X^2_{ij}X_{i+}X_{+j}$
11	$(\sum_{i=1}^{n} X^2_{i+})(\sum_{j=1}^{k} X^2_{+j})$	27	$(Y_{++})^2$
12	$(X_{++})^2 Y_{++}$	28	$(X_{++})(Z_{++})$
13	$\sum_{j=1}^{k}(\sum_{i=1}^{n} X_{ij}X_{i+})^2$	29	$\sum_{i=1}^{n} Y^2_{i+}$
14	$\sum_{i=1}^{n}(\sum_{j=1}^{k} X_{ij}X_{+j})^2$	30	$\sum_{j=1}^{k} Y^2_{+j}$
15	$\sum_{i=1}^{n}(\sum_{j=1}^{k} X_{ij}X_{i+})(\sum_{j=1}^{k} X_{ij}X_{+j})$	31	$\sum_{i=1}^{n} X_{i+}Z_{i+}$
16	$\sum_{j=1}^{k}(\sum_{i=1}^{n} X_{ij}X_{i+})(\sum_{i=1}^{n} X_{ij}X_{+j})$	32	$\sum_{j=1}^{k} X_{+j}Z_{+j}$
		33	$\sum_{i=1}^{n}\sum_{j=1}^{k} X^4_{ij}$

Table 3-8. Conversion of Sigmas (Σ's) to f-Statistics

		Σ														
f	*Constant*	*1*	*2*	*3*	*4*	*5*	*6*	*7*	*8*	*9*	*10*	*11*	*12*	*13*	*14*	*15*
1	$1/(n^{[4]}k^{[4]})$	1	−6	−6	3	3	8	8	−6	−6	24	6	6	−12	−12	−24
2	$1/(n^{[3]}k^{[4]})$		1	0	−1	0	−2	0	2	0	−4	−1	−1	4	2	6
3	$1/(n^{[4]}k^{[3]})$			1	0	−1	0	−2	0	2	−4	−1	−1	2	4	4
4	$1/(n^{[2]}k^{[4]})$				1	0	0	0	−1	0	0	0	0	−4	0	0
5	$1/(n^{[4]}k^{[2]})$					1	0	0	0	−1	0	0	0	0	−4	0
6	$1/(n^{[2]}k^{[4]})$						1	0	−1	0	0	0	0	0	0	−3
7	$1/(n^{[4]}k^{[2]})$							1	0	−1	0	0	0	0	0	0
8	$1/(n\quad k^{[4]})$								1	0	0	0	0	0	0	0
9	$1/(n^{[4]}k\quad)$									1	0	0	0	0	0	0
10	$1/(n^{[3]}k^{[3]})$										1	0	0	−1	−1	−1
11	$1/(n^{[3]}k^{[3]})$											1	0	0	0	−2
12	$1/(n^{[3]}k^{[3]})$												1	0	0	0
13	$1/(n^{[2]}k^{[3]})$													1	0	0
14	$1/(n^{[3]}k^{[2]})$														1	0
15	$1/(n^{[2]}k^{[3]})$															1
16	$1/(n^{[3]}k^{[2]})$															
17	$1/(n^{[2]}k^{[2]})$															
18	$1/(n^{[2]}k^{[3]})$															
19	$1/(n^{[3]}k^{[2]})$															
20	$1/(n^{[2]}k^{[3]})$															
21	$1/(n^{[3]}k^{[2]})$															
22	$1/(n\quad k^{[3]})$															
23	$1/(n^{[3]}k\quad)$															
24	$1/(n^{[2]}k^{[2]})$															
25	$1/(n^{[2]}k^{[2]})$															
26	$1/(n^{[2]}k^{[2]})$															
27	$1/(n^{[2]}k^{[2]})$															
28	$1/(n^{[2]}k^{[2]})$															
29	$1/(n\quad k^{[2]})$															
30	$1/(n^{[2]}k\quad)$															
31	$1/(n\quad k^{[2]})$															
32	$1/(n^{[2]}k\quad)$															
33	$1/(n\quad k\quad)$															

Σ

16	*17*	*18*	*19*	*20*	*21*	*22*	*23*	*24*	*25*	*26*	*27*	*28*	*29*	*30*	*31*	*32*	*33*
−24	6	−6	−6	−24	−24	36	36	24	24	48	3	16	−18	−18	−48	−48	36
4	−2	2	1	6	4	−12	−6	−6	−8	−12	−1	−4	6	6	16	12	−12
6	−2	1	2	4	6	−6	−12	−8	−6	−12	−1	−4	6	6	12	16	−12
0	2	−2	0	0	0	6	0	0	8	0	1	0	−3	−6	−8	0	6
0	2	0	−2	0	0	0	6	8	0	0	1	0	−6	−3	0	−8	6
0	0	0	0	−3	0	6	0	3	0	6	0	2	−3	0	−8	−6	6
−3	0	0	0	0	−3	0	6	0	3	6	0	2	0	−3	−6	−8	6
0	0	0	0	0	0	−6	0	0	0	0	0	0	3	0	8	0	−6
0	0	0	0	0	0	0	−6	0	0	0	0	0	0	3	0	8	−6
−1	1	0	0	−1	−1	2	2	2	2	3	0	1	−2	−2	−4	−4	4
−2	0	−1	−1	0	0	2	2	2	2	4	1	0	−2	−2	−4	−4	4
0	0	−1	−1	−2	−2	2	2	2	2	2	1	2	−2	−2	−4	−4	4
0	−1	0	0	0	0	−1	0	0	−2	0	0	0	1	2	2	0	−2
0	−1	0	0	0	0	0	−1	−2	0	0	0	0	2	1	0	2	−2
0	0	0	0	0	0	−1	0	−1	0	−2	0	0	1	0	2	2	−2
1	0	0	0	0	0	0	−1	0	−1	−2	0	0	0	1	2	2	−2
	1	0	0	0	0	0	0	0	0	0	0	0	−1	−1	0	0	1
		1	0	0	0	−1	0	0	−2	0	−1	0	1	2	2	0	−2
			1	0	0	0	−1	−2	0	0	−1	0	2	1	0	2	−2
				1	0	−1	0	−1	0	−1	0	−1	1	0	2	2	−2
					1	0	−1	0	−1	−1	0	−1	0	1	2	2	−2
						1	0	0	0	0	0	0	−1	0	−2	0	2
							1	0	0	0	0	0	0	−1	0	−2	2
								1	0	0	0	0	−1	0	0	−1	1
									1	0	0	0	0	−1	−1	0	1
										1	0	0	0	0	−1	−1	1
											1	0	−1	−1	0	0	1
												1	0	0	−1	−1	1
													1	0	0	0	−1
														1	0	0	−1
															1	0	−1
																1	−1
																	1

Table 3-9. Computation of F-Statistics from f-Statistics

	f															
F	*1*	*2*	*3*	*4*	*5*	*6*	*7*	*8*	*9*	*10*	*11*	*12*	*13*	*14*	*15*	*16*
1	1															
2	−1	1														
3	−1	0	1													
4	1	−2	0	1												
5	1	0	−2	0	1											
6	2	−3	0	0	0	1										
7	2	0	−3	0	0	0	1									
8	−6	12	0	−3	0	−4	0	1								
9	−6	0	12	0	−3	0	−4	0	1							
10	1	−1	−1	0	0	0	0	0	0	1						
11	1	−1	−1	0	0	0	0	0	0	0	1					
12	1	−1	−1	0	0	0	0	0	0	0	0	1				
13	−1	2	1	−1	0	0	0	0	0	−2	0	0	1			
14	−1	1	2	0	−1	0	0	0	0	−2	0	0	0	1		
15	−2	3	2	0	0	−1	0	0	0	−2	−1	0	0	0	1	
16	−2	2	3	0	0	0	−1	0	0	−2	−1	0	0	0	0	1
17	1	−2	−2	1	1	0	0	0	0	4	0	0	−2	−2	0	0
18	−1	2	1	−1	0	0	0	0	0	0	−1	−1	0	0	0	0
19	−1	1	2	0	−1	0	0	0	0	0	−1	−1	0	0	0	0
20	−2	3	2	0	0	−1	0	0	0	−2	0	−1	0	0	0	0
21	−2	2	3	0	0	0	−1	0	0	−2	0	−1	0	0	0	0
22	6	−12	−6	3	0	4	0	−1	0	8	2	2	−2	0	−2	0
23	6	−6	−12	0	3	0	4	0	−1	8	2	2	0	−2	0	−2
24	2	−3	−4	0	2	1	0	0	0	4	1	1	0	−2	−1	0
25	2	−4	−3	2	0	0	1	0	0	4	1	1	−2	0	0	−1
26	4	−6	−6	0	0	2	2	0	0	6	2	1	0	0	−2	−2
27	1	−2	−2	1	1	0	0	0	0	0	2	2	0	0	0	0
28	4	−6	−6	0	0	2	2	0	0	6	0	3	0	0	0	0
29	−6	12	12	−3	−6	−4	0	1	0	−16	−4	−4	4	8	4	0
30	−6	12	12	−6	−3	0	−4	0	1	−16	−4	−4	8	4	0	4
31	−12	24	18	−6	0	−8	−6	2	0	−24	−6	−6	6	0	6	6
32	−12	18	24	0	−6	−6	−8	0	2	−24	−6	−6	0	6	6	6
33	36	−72	−72	18	18	24	24	−6	−6	96	24	24	−24	−24	−24	−24

Equations (3.1) through (3.4) may be used with *all* item-scoring procedures and are not restricted to cases where items are scored dichotomously. In deriving these equations, it was assumed that both the population and universe were finite; also, no restrictions were placed on n and k. Equations related to (3.1) through (3.4) have been derived by Lord (1960) for the case when items are scored dichotomously and the value of n is reasonably large (i.e., 500). The equations given in this chapter are more generally applicable and permit multiple matrix sampling being used with a wider variety of sampling plans.

Understanding the computations involved in Equations (3.1) through (3.4) is facilitated by expressing them in terms of statistics used more commonly. For example, Equation (3.1) is an identify for

j																
17	*18*	*19*	*20*	*21*	*22*	*23*	*24*	*25*	*26*	*27*	*28*	*29*	*30*	*31*	*32*	*33*
1																
0	1															
0	0	1														
0	0	0	1													
0	0	0	0	1												
0	−1	0	−2	0	1											
0	0	−1	0	−2	0	1										
0	0	−1	−1	0	0	0	1									
0	−1	0	0	−1	0	0	0	1								
0	0	0	−1	−1	0	0	0	0	1							
0	−2	−2	0	0	0	0	0	0	0	1						
0	0	0	−3	−3	0	0	0	0	0	0	1					
−2	2	4	4	0	−2	0	−4	0	0	−1	0	1				
−2	4	2	0	4	0	−2	0	−4	0	−1	0	0	1			
0	3	0	6	6	−3	0	0	−3	−3	0	−1	0	0	1		
0	0	3	6	6	0	−3	−3	0	−3	0	−1	0	0	0	1	
6	−12	−12	−24	−24	12	12	12	12	12	3	4	−3	−3	−4	−4	1

$$\hat{\mu}'_{1_S} = \frac{K}{k}\,\overline{T} \tag{3.1a}$$

and Equation (3.2) may be reduced to

$$\hat{\mu}_{2_S} = \frac{nK\left\{(K-1)s_T^2 - (K-k)\sum_{i=1}^{k} v_i\right\}}{k(k-1)(n-1)} \tag{3.2a}$$

where,

s_T^2 refers to the variance of subtest scores,

$\overline{T}$ to the mean subtest score, and

v_i to the variance of item scores for item i.

Suffice it to say that the estimators defined by Equations (3.3) and (3.4) are functions of the moments of the subtest scores, moments of the item difficulty indices, the inter-item correlations, and the correlation between each item and the total subtest score.

Estimation of Reliability and Components of Variance

The reliability of universe scores is estimated through an analysis of variance procedure derived by Winer (1971). In this procedure, the average reliability of a single item in the subtest is estimated and then inserted into the well-known Spearman-Brown Prophesy Formula to obtain the reliability of scores on the K-item test or universe. The reliability coefficient computed in this manner is equal to

$$\hat{r}_{KK} = \frac{K\hat{\theta}'}{1 + K\hat{\theta}'} \qquad (3.5)$$

where,

$$\hat{\theta}' = \frac{\hat{MS}_E - m\hat{MS}_{IE}}{mk\,\hat{MS}_{IE}}, \qquad (3.6)$$

MS_E = mean square between examinees,

MS_{IE} = residual mean square, and

$m = [n(k-1) - 2] / [n(k-1)]$

It should be noted here that θ' is not an estimate of the average item reliability (r_{11}) but is, instead, a function of r_{11}. Specifically,

$$\hat{r}_{11} = \frac{\hat{\theta}'}{1 + \hat{\theta}'} .$$

If this identity were substituted into the Spearman-Brown Prophesy Formula, Equation (3.5) would be the result.

The components of variance for items, examinees, and residual are also estimated easily by

$$\hat{\sigma}_I^2 = \frac{\hat{MS}_I - \hat{MS}_{IE}}{n}, \tag{3.7}$$

$$\hat{\sigma}_E^2 = \frac{\hat{MS}_E - \hat{MS}_{IE}}{k}, \tag{3.8}$$

and

$$\hat{\sigma}_{IE}^2 = \hat{MS}_{IE}. \tag{3.9}$$

The expected mean squares used in computing the components of variance defined by (3.7), (3.8), and (3.9) are those obtained using the additive model defined by Myers (1966). These computations are not unique and are found in numerous sources. Equation (3.7) is of particular interest because it estimates the variance of the item difficulty indices. One direction in which multiple matrix sampling can be expected to expand in the future is towards the use of more than two facets (i.e., more than *items* and *examinees*) in partitioning the variance of subtest scores. The framework for this expansion has been provided by Cronbach, Gleser, Nanda, and Rajaratnam (1972), Bennett (1956), and Carney (1968).

Alternative Procedure for Estimating Moments

Equations (3.1) through (3.4) define the simplest procedures for estimating moments of test scores over the item universe. An additional procedure has been reported by Ray, Hundleby, and Goldstein (1962). It is known commonly that the mean, variance, and third and fourth central moments of total scores on the item universe can be written as functions of item statistics. For example,

$$\mu_1' = \sum_{j=1}^{K} p_j \tag{3.10}$$

and

$$\mu_2 = \sum_{j=1}^{K} \sigma_j^2 + \sum_{\substack{i=1 \\ i \neq j}}^{K} \sum_{j=1}^{K} \sigma_{ij} \tag{3.11}$$

where

p_j refers to the mean item score for item j,

σ_j^2 to the variance of item scores on item j, and

σ_{ij} to the covariance of scores on item i with those on item j.

Similar equations are given by Ray, Hundleby, and Goldstein for estimating μ_3 and μ_4 for the case where items are scored dichotomously and involve such terms as p_{ijk} and p_{ijkl}. The first term refers to the proportion of examinees answering correctly item i, item j, *and* item k; the second, to the proportion of examinees answering correctly item i, item j, item k, *and* item l. To estimate moments of test score distributions in this procedure, several requirements must be met: (a) each item in the universe must be included in some subtest, (b) each of the $K(K-1)/2$ possible item pairing must be included in some subtest, (c) each of the $K(K-1)(K-2)/6$ item triplets must be tested, as well as, (d) each of the $K(K-1)(K-2)(K-3)/24$ item quadruples. To meet these requirements, an extremely large number of subtests must be used. From a practical viewpoint the procedure is feasible only for very small item universes—probably those having fewer than 50 items. For this reason, the procedure is merely referenced here.

CALCULATION OF POOLED ESTIMATES OF PARAMETERS

In multiple matrix sampling it is not necessary that each subtest have the same number of items or that the same number of examinees respond to each subtest. The differences among the numbers of items in each subtest and examinees per subtest are taken into consideration in computing the pooled or combined estimate of each parameter. The weighted mean of the estimates for each parameter over subtests is used to obtain the best single estimate of the parameter. Specifically,

$$\hat{\mu}_p = \frac{\sum_{s=1}^{t} 0_s \hat{\mu}_s}{\sum_{s=1}^{t} 0_s} \tag{3.12}$$

where μ refers to any parameter, t, to the number of subtests and $O_s = n_s k_s$, the number of observations obtained from subtest s.

Although there are numerous procedures by which the t estimates of each parameter could be pooled into a composite, Equation (3.12) is used most commonly. Alternative procedures for combining individual estimates of a particular parameter have not been examined systematically. It should be noted that Equation (3.12) is a component of the jackknife procedure discussed at a later point in this chapter.

ESTIMATING PARAMETERS OF SUBUNIVERSES

Using Equations (3.1) through (3.9) the investigator may estimate readily the parameters of the total scores over the item universe. It is often the case that the item universe may be stratified (by design or by accident) according to some variable of interest. Often, the total score (on entire universe) is quite meaningless; what is more meaningful and more important is estimating the parameters of the total scores on each stratum of interest within the universe. In this case, Equations (3.1) through (3.9) are applied to each stratum and K refers to the total number of items within the particular stratum (subuniverse) of interest. The entire multiple matrix sampling procedure is repeated for each stratum of interest and all the guidelines discussed in Chapter 2 are applicable. Consider the example in which a 600-item universe has been constructed with 300 of the items being in English and 300 in Spanish. Furthermore, the 300 items in each language may be further subdivided into 100 mathematics items and 200 social studies items. Here, a total score on the 600-item universe is uninterpretable. Meaningful universe score distributions would be those for English language mathematics items, English language social studies items, Spanish language mathematics items, and Spanish language social studies items. For each of these four content strata the entire multiple matrix sampling procedure must be used. If the investigator plans to estimate normative distributions for substrata, a stratified-random sampling plan must be used for assigning items within each content stratum to individual subtests.

ESTIMATING STANDARD ERRORS OF ESTIMATE

Computing the pooled estimate of each parameter is only solving half of the problem. Of equal importance is estimating the standard error of estimate associated with this value. Two general procedures are available for estimating the standard error of estimate. The first is one outlined originally by Hooke (1956a, 1956b) and used by Lord and Novick (1968); the second, known as the "jackknife" procedure, has been popularized by Mosteller and Tukey (1968). Although both procedures may be used when items are scored polychotomously, the first may be used only when the items for each subtest have been sampled without replacement from the universe. The jackknife procedure may be used with *all* item-sampling plans (i.e., with replacement and without replacement.)

The Hooke-Lord-Novick Equations

If items and examinees have been sampled randomly and without replacement for each subtest, the squared standard error of the estimate of the mean universe score for subtest s is equal to

$$\underset{N \to \infty}{VAR}(\hat{\mu}'_{1_s}) = \left\{ \frac{1}{nk} D_4 + \frac{1}{n} D_2 + \left(\frac{1}{k} - \frac{1}{K} \right) D_3 \right\} K^2 \qquad (3.13)$$

The D-statistics are computed using Tables 3-1, 3-2, and 3-3. Equation (3.13) estimates the error variance associated with the estimate of the mean universe score obtained from *one* subtest. The estimate of the standard error of the *pooled* estimate of the mean, however, is a function of (3.13) and is computed as

$$\underset{N \to \infty}{SE}(\hat{\mu}'_{1_p}) = \frac{1}{t} \left\{ \sum_{s=1}^{t} VAR(\hat{\mu}'_{1_s}) - \frac{(t-1)K^2 \sum_{s=1}^{t} \hat{\sigma}^2_{I_s}}{(K-1)} \right\}^{1/2} \qquad (3.14)$$

where $\sigma^2_{I_s}$ refers to the variance of the mean item scores for subtest s. Equation (3.14) may be used only when the items for each subtest have been sampled without replacement from the universe.

A variation of (3.14) has been given by Lord and Novick (1968) for the case when items are scored dichotomously. Their equation when modified to give the standard error of the mean universe score (instead of the mean proportion correct score) is

$$SE(\hat{\mu}'_{1_p}) = \left\{ \left[\frac{1}{tkn}\right] \left[\frac{1}{(K-1)(N-1)}\right] \left[K^2 N \hat{\sigma}^2_{I_p} \left\{ (K-k)(n-1) - kn(t-1) \right\} + K \hat{\mu}_{2_p} \left\{ (N-n)(k-1) - kn(t-1) \right\} + \hat{\mu}'_{1_p}(K - \hat{\mu}'_{1_p}) \left\{ (K-k)(N-n) + kn(t-1) \right\} \right] \right\}^{1/2} \qquad (3.15)$$

In most applications, the population of interest is large. Modifying Equation (3.15) for this case produces

$$\underset{N\to\infty}{SE(\hat{\mu}'_{1p})} = \left\{ \left[\frac{1}{tkn}\right] \left[\frac{1}{(K-1)}\right] \left[K^2 \sigma^2_{I_p} \left\{(K-k)(n-1) - kn(t-1)\right\} \right.\right. \tag{3.15a}$$

$$\left.\left. + K\hat{\mu}_{2p}(k-1) + \hat{\mu}'_{1p}(K-\hat{\mu}'_{1p})(K-k) \right] \right\}^{1/2}$$

which is equivalent to (3.14) when items are scored dichotomously.

The squared standard error of the estimate of the variance of universe scores using the results of subtest s is equal to

$$\begin{aligned}
\underset{N\to\infty}{VAR(\hat{\mu}_{2s})} = & \left[\frac{1}{n}\left\{F_8 + \frac{2}{K}F_{22} + \frac{1}{K^2}F_{29}\right\}\right. \\
& + \left(\frac{1}{k} - \frac{1}{K}\right)\left\{4F_{13} + \frac{4}{K}F_{25} + \frac{1}{K^2}F_{30}\right\} \\
& + \frac{2}{n-1}\left\{F_4 + \frac{2}{K}F_{13} + \frac{1}{K^2}F_{17}\right\} \\
& + \frac{2}{K^2}\left(\frac{1}{k-1} - \frac{1}{K-1}\right)F_{17} \\
& + \left(1 - \frac{2}{K}\right)\left\{\frac{2}{k(k-1)} - \frac{2}{K(K-1)}\right\}F_{17} \qquad (3.16) \\
& + \frac{2}{k(n-1)}\left\{2(F_{13} + F_{18}) + \frac{4}{K}F_{25}\right. \\
& \left. + \frac{1}{K^2}F_{30}\right\} + \frac{2}{K^2}\left\{\frac{1}{n(k-1)}F_{29}\right\} \\
& + \frac{1}{nk}\left\{4F_{22} + \frac{4}{K}F_{31} + \frac{1}{K^2}F_{33}\right\} \\
& + \frac{1}{K^2}\left\{\frac{2}{(n-1)(k-1)}\right\}(F_{17} + F_{27})
\end{aligned}$$

$$+\left(1-\frac{2}{K}\right)\left\{\frac{2}{nk(k-1)}\right\}F_{29}$$

$$+\left(1-\frac{2}{K}\right)\left\{\frac{2}{k(n-1)(k-1)}\right\}(F_{17}+F_{27})\Bigg]K^4 \tag{3.16}$$

Tables 3-7, 3-8, and 3-9 are used to compute the F-statistics required in (3.16). The standard error of the pooled estimate of the variance of universe scores is computed as follows

$$\underset{N\to\infty}{SE(\hat{\mu}_{2_p})} = \frac{1}{t}\left\{\sum_{s=1}^{t} VAR(\hat{\mu}_{2_s}) - (t-1)4K\sum_{s=1}^{t} VAR(s_{kz})_s\right\}^{1/2} \tag{3.17}$$

where VAR(s_{kz})$_s$ refers to the variance of the covariances of each item with the mean item scores for subtest s. Equation (3.17) is applicable only when items have been sampled without replacement for each subtest. Equations comparable to (3.13) and (3.16) could be derived for $\hat{\mu}_3$ and $\hat{\mu}_4$ within the framework developed by Hooke; however, doing so would not be a simple undertaking. In Hooke's procedure, estimating the standard error of the ith moment requires estimating the 2 ith moment and the number of sums (Σ's in Tables 3-1, 3-4, and 3-7) becomes large. For example, for the 5th, 6th, 7th, and 8th moment, 91, 298, 910, and 3017 sums, respectively, would be computed. Were one to pursue this line of research, the work of Dayhoff (1964, 1966) will be of use.

An alternative procedure—called the "jackknife"—exists for estimating standard errors of estimate and is described next. The jackknife procedure could be used to compute the standard error of the pooled estimates of the third and fourth central moments after the results from each subtest had been collected. One advantage of the jackknife over Equations (3.14) and (3.17) is that the jackknife procedure may be used to compute the standard error of the pooled estimate for all item-sampling plans (i.e., with replacement and without replacement). The jackknife procedure, like Equation (3.14), has a standard error of estimate associated with it because in each case the estimator contains terms which are estimates of parameters. Determining the standard error of the procedure used to estimate the standard error is important; however, little has been done in this area.

The Jackknife Procedure

The jackknife procedure, described originally by Quenouille (1956) as a method for bias reduction in estimating parameters, is a rather uncomplicated procedure which may be used in multiple matrix sampling to approximate the standard error of estimate associated with numerous estimates of parameters. Although there appear to be a few situations (see, for example, Collins, 1970) where it does not work satisfactorily, it does perform quite well in a surprisingly large number of situations. The name *jackknife* was given to this procedure by Tukey (Mosteller & Tukey, 1968) to "suggest the broad usefulness of the technique as a substitute for specialized tools that may not be available, just as the Boy Scout's trusty tool serves so variedly." A good description of the jackknife applied to a variety of estimation problems is given by Mosteller and Tukey (1968) and additional descriptions are given by Miller (1964), Jones (1965), and Mosteller (1971).

The jackknife operates on a data set which has been divided into subgroups of data and gives a mean estimate of the parameter computed over subgroups and an estimate of the standard error of estimate associated with this estimate. A basic component of the jackknife is the pseudovalue associated with each subgroup which, for each subgroup, is the weighted difference between the statistic computed on all the data and the statistic computed on the body of data that remains after omitting that subgroup. Because the pseudovalues "behave as though they are independent of each other," the standard error of the statistic is computed according to the well-known formula for the standard error of a sample mean, $s_{\overline{X}} = s_X/N^{1/2}$. The computations are relatively simple. Let

t the number of subgroups,
y_{all} the statistic computed on all the data, and
$y_{(j)}$ the statistic computed on all the data left after removing subgroup j.

The pseudovalues, y_{*j}, are then equal to

$$y_{*j} = t\,y_{all} - (t-1)\,y_{(j)} \text{ for } j = 1, 2, \ldots, t. \tag{3.18}$$

The jackknifed estimate of the parameter is equal to

$$y_{*} = (y_{*1} + y_{*2} + \ldots + y_{*t})/t \tag{3.19}$$

with an estimate of its variance given by

$$s_{*}^{2} = \frac{\sum_{j=1}^{t} (y_{*j} - y_{*})^2}{t(t-1)} \tag{3.20}$$

When the jackknife is applied to multiple matrix sampling, there are t subgroups of data (subtests) but only one score for each subgroup (the estimate of the parameter) with that score weighted according to the number of observations nk acquired through that subtest. If the statistics computed on each subgroup are weighted equally, the pseudovalues reduce algebraically to the means for the subgroups. In this case y_* is equal to y_{all} and s_*^2 is equal to the variance of the subgroup statistics. It should be noted that the jackknife is most appropriate when the items for each subtest are sampled randomly from the universe, that is, after the items for one subtest have been selected, these items are not removed from the universe when selecting items for another subtest. If items for each subtest are sampled randomly from the universe but without replacement, the jackknife procedure approximates conservatively the standard error of estimate. Shoemaker (1973) has verified empirically this conclusion.

RESULTS IN ESTIMATING SPELLING PROFICIENCY

The results from the Shoemaker and Okada investigation of spelling proficiency are given in Tables 3-10 and 3-11. The estimates of parameters in Table 3-10

Table 3-10. Estimates of Moments in Shoemaker and Okada Investigation

Subtest	k	n	$\hat{\mu}'_1$	$\hat{\mu}_2$	$\hat{\mu}_3$	$\hat{\mu}_4$
1	10	18	17.778	156.51	−25.55	38881.
2	10	14	26.429	60.06	219.91	−2460.
3	10	13	23.462	233.08	603.30	66050.
4	10	13	24.615	271.09	361.45	114130.
5	10	12	9.583	166.09	1547.10	21412.
Mean			20.417	173.55	465.46	46351.
SE			2.105	24.47	260.99	18796.

Table 3-11. Components of Variance for Shoemaker and Okada Investigation

Subtest	k	n	$\hat{\sigma}^2_{\text{I}}$	$\hat{\sigma}^2_{\text{E}}$	$\hat{\sigma}^2_{\text{IE}}$	$\hat{\theta}'$
1	10	18	.0218	.0602	.1535	.3987
2	10	14	.1000	.0187	.1429	.1348
3	10	13	.0207	.0890	.1494	.6089
4	10	13	.0397	.1039	.1192	.8893
5	10	12	−.0002	.0658	.0955	.7053
Mean			.0368	.0660	.1349	.5236
SE*			.0167	.0140	.0105	.1270

*All standard errors were calculated using equation (3.20)

resulted from Equations (3.1) through (3.4); those in Table 3-11, from Equations (3.6) through (3.9). The mean estimate for each parameter resulted from Equation (3.12); the standard error of estimate, from Equations (3.14) and (3.17) for the mean and variance, and Equation (3.20) for the third and fourth central moments. Using Equation (3.5), the reliability of the universe scores was found to be .963. Use of the jackknife in computing the pooled estimate of each parameter and the standard error of estimate for each parameter is demonstrated easily for the case of the mean universe score. For this data (using the column headed $\hat{\mu}'_1$ in Table 3-9),

$$\hat{\mu}'_{1all} = \frac{\sum_{s=1}^{t} 0_s \hat{\mu}'_{1s}}{\sum_{s=1}^{t} 0_s} = \frac{14300.0900}{700} = 20.4287$$

$$\hat{\mu}'_{1(1)} = \frac{\sum_{s=1}^{t} 0_s \hat{\mu}'_{1s} - 0_1 \hat{\mu}'_{11}}{\sum_{s=1}^{t} 0_s - 0_1}$$

$$= \frac{14300.0900 - (180)(17.7780)}{700 - 180} = 21.3462$$

$$\hat{\mu}'_{1(2)} = 18.9287$$

$$\hat{\mu}'_{1(3)} = 19.7370$$

$$\hat{\mu}_{1(4)} = 19.4738, \text{ and}$$

$$\hat{\mu}_{1(5)} = 22.6725.$$

The computed pseudovalues are equal to

$$\hat{\mu}'_{1*1} = t\hat{\mu}'_{1all} - (t-1)\hat{\mu}'_{11}$$

$$= (5)(20.4287) - (4)(21.3462) = 16.7587$$

$$\hat{\mu}'_{1*2} = 26.4287$$

$$\hat{\mu}'_{1*3} = 23.1955$$

$$\hat{\mu}'_{1*4} = 24.2483, \text{ and}$$

$$\hat{\mu}'_{1*5} = 11.4535.$$

The pooled estimate of the parameter is

$$\hat{\mu}'_{1*} = (16.7587 + 26.4287 + \ldots + 11.4535)/5 = 20.4169$$

Its associated standard error of estimate is the square root of 7.66133 or 2.7592.

COMPUTING CONFIDENCE INTERVALS FOR PARAMETERS

The confidence interval for a given parameter is a numerical interval computed in such a manner that the true value of the parameter lies within the interval with a given probability. For example, if the 95% confidence interval for the mean universe score was computed to be 28 to 32, the probability that the true mean universe score lies within this interval is .95. When computing a confidence interval in multiple matrix sampling, the investigator must have available (a) a pooled estimate of the parameter, (b) an estimate of the standard error associated with this pooled estimate, and (c) knowledge of the theoretical sampling distribution of the pooled estimate. The difficulty, however, with computing confidence intervals in multiple matrix sampling is that the theoretical sampling distribution of the pooled estimate has been investigated for only a few parameters. Sirotnik (1973) has approximated empirically the theoretical sampling distribution of the mean proportion correct score using one multiple matrix sampling plan and Shoemaker (1973) has done the same for the mean universe score, variance of universe scores, and third and fourth central moments of universe scores using a variety of multiple matrix sampling plans. The combined results of these two preliminary investigations suggest that the sampling distribution of the pooled estimate is a function of the distribution of universe scores in the population of examinees. Specifically, when the universe scores are distributed normally, the sampling distribution of the pooled estimate of the mean universe score, variance, and third central moment of universe scores is normal and that for the fourth central moment is distributed according to the chi-square distribution with the degrees of freedom one less than the number of subtests. When the universe scores are *not* distributed normally in the population, the sampling distributions of these estimates follow other distributions as yet undetermined. However, in this latter case, the sampling distribution of the pooled estimate of the mean universe score and variance of universe scores are approximately normal and confidence intervals constructed upon this assumption should be adequate for most purposes. Additional research is required to clarify these tentative conclusions.

The procedure for computing the $(1 - \alpha)$ confidence interval for a parameter P assuming that the sampling distribution of the pooled estimate is normal is

$$\hat{P}_p - z_{(\alpha/2)} SE(\hat{P}_p) \angle P \angle \hat{P}_p + z_{(1-\alpha/2)} SE(\hat{P}_p) \qquad (3.21)$$

where $\hat{P}_p$ is the pooled estimate of the parameter, and $z_{(\alpha/2)}$ and $z_{(1-\alpha/2)}$ are the values in the *ND* (0,1) normal distribution corresponding respectively to the $\alpha/2$ and $(1 - \alpha/2)$ percentiles. Using the results reported in Table 3-10 as an example, the .95 confidence interval ($\alpha = .05$) for the mean universe score is computed as

$$20.417 - 1.960(2.105) \angle \mu'_1 \angle 20.417 + 1.960(2.105)$$

$$16.291 \angle \mu'_1 \angle 24.523$$

If the chi-square distribution were used to calculate the confidence interval for a given parameter, the corresponding percentiles in the chi-square distribution having $(t - 1)$ degrees of freedom would be substituted into the following inequality

$$\hat{P}_p + \frac{\left[\chi^2_{(\alpha/2)} - df\right] SE(\hat{P}_p)}{(2df)^{1/2}} \angle P \angle \hat{P}_p + \frac{\left[\chi'^2_{(1-\alpha/2)} - df\right] SE(\hat{P}_p)}{(2df)^{1/2}}. \qquad (3.22)$$

The sampling distributions of pooled estimates in multiple matrix sampling have only begun to be examined and a more thorough investigation of these distributions can be expected in the future.

COMPUTATIONAL IRREGULARITIES

In estimating a parameter from subtest results it sometimes happens that the estimate of a parameter which is inherently positive will be negative. An example of this irregularity is found in both Table 3-10 and Table 3-11. In Table 3-10, the estimate of μ_4 using subtest 2 was negative and, in Table 3-11, the component of variance for examinees estimated from subtest 5 was negative. Components of variance are inherently positive as are also the variance and fourth central moment of the distribution of test scores. Although uninterpretable, estimates such as these should *not* be discarded or set equal to zero in computing the pooled estimate of the parameter. It must be remembered that the results of any subtest are relatively unimportant; what is important is the accuracy of the pooled estimate of the parameter. Any procedure which ignores part of the data produces an estimate of the parameter which is biased, i.e., it would not approach the true value even if the number of subtests was increased indefinitely. Sirotnik (1970) has verified empirically this conclusion regarding negative estimates of the variance of universe scores. The point to be made is that these negative estimates should be left alone.

APPROXIMATING THE NORMATIVE DISTRIBUTION

In addition to estimating individual parameters through multiple matrix sampling, it is possible to estimate the entire normative frequency distribution of test scores which would have been obtained by testing all examinees on all items in the universe of interest. A variety of distributions may be used to approximate the normative distribution but determining the most appropriate distribution has received little attention in the literature.

Testing Feasibility of Estimated Moments

All of the graduation procedures to be described require estimates of the moments of the distribution of universe scores. Because the moments of a distribution are interrelated, i.e., functions of one another, the limits of the $k+1$ moment (the values which it can assume) are determined by the first k moments. It sometimes happens in multiple matrix sampling that the pooled estimates of the third and/or fourth central moments are not consistent with the lower order moments. In this case, no distribution exists which has those moments and any graduation procedure attempted will fail. When this happens, the graduation procedure selected must be based solely upon those moment estimates which are within the accepted limits or the moment estimates which are not feasible must be modified. One procedure (and the one incorporated into the computer program in Appendix A) for assessing the feasibility of the estimated second through fourth central moments is the linear programming technique discussed in detail by Hill (1972). Applying Hill's linear programming technique to the pooled moment estimates reported in Table 3-10 produces the minimum and maximum values for each moment given in Table 3-12. In this case, all estimates are within limits, which means that at least one distribution exists having these moments. This procedure does not indicate which distribution is to be used but indicates merely that a distribution does exist. The feasibility of the estimated moments should be determined before any graduation procedure is attempted.

The Negative Hypergeometric Distribution

The negative hypergeometric distribution has been shown by Keats and Lord (1962) to provide a reasonably good fit for a wide variety of test score distributions when the test score is the number of correct responses (as would be the case if the items were scored dichotomously; 1 = pass, 0 = fail). The negative hypergeometric distribution is a function of the mean universe score, variance of universe scores, and the number of items in the universe. Lord (1962) and Shoemaker (1970) have demonstrated that the negative hypergeometric distribution with parameters estimated by multiple matrix sampling can be used satisfactorily to approximate normative distributions of number correct scores. The procedure used by them in arriving at this conclusion was postmortem item-examinee sampling—item-examinee samples are taken from an existing data base (obtained by actually testing all examinees over all items) and used to estimate the parameters of the data base from which they have been sampled. In this procedure, the true distribution of universe scores is known in advance and each estimated distribution based upon estimates obtained through multiple matrix sampling may be compared to it.

The formula for the negative hypergeometric distribution is

$$h(T) = \frac{c(-K)_T(a)_T}{(-b)_T T!} \quad \text{for } T = 0, 1, 2, \ldots, K \tag{3.23}$$

Table 3-12. Minimum and Maximum Values for Estimated Moments

	About Origin			About Mean		
Moment	*Estimate*	*Minimum*	*Maximum*	*Estimate*	*Minimum*	*Maximum*
1	20.417			0.000		
2	590.389	417.082	1020.832	173.550	.243	603.993
3	19605.840	17073.870	23249.080	465.465	−2066.512	4108.699
4	692172.400	677289.700	751954.600	46350.920	31467.250	106132.100

where

$$a = (-1 + 1/\hat{\alpha}_{21})\hat{\mu}_{1p}$$

$$\hat{\alpha}_{21} = \frac{K}{K-1}\left[1 - \frac{\hat{\mu}'_{1p} - \hat{\mu}'^{2}_{1p}/K}{\hat{\mu}_{2p}}\right]$$

$$b = -a - 1 + \frac{K}{\hat{\alpha}_{21}}$$

$$c = \frac{b^{[K]}}{(a+b)^{[K]}}$$

$$b^{[K]} = b(b-1)(b-2)\ldots(b-K+1)$$

$$(a)_T = a(a+1)(a+2)\ldots(a+T-1)$$

$$(a)_0 = b^{[0]} = 1, \text{ and}$$

$$T! = T(T-1)(T-2)\ldots(2)(1).$$

The calculations involved in the negative hypergeometric distribution have been incorporated into the computer program given in Appendix A. Using the pooled estimate of the mean universe score and variance of scores obtained in the Shoemaker and Okada investigation, the relative frequency per score computed by Equation (3.23) are given in Table 3-13. These relative frequencies are the individual probabilities associated with all possible universe scores. For example, the probability of an examinee spelling correctly 20 words out of 50 words is .024. An equally appropriate interpretation is that 2.4% of the population can spell correctly 20 words. As should be the case, the relative frequencies in Table 3-13 sum to unity. An estimate of the number of examinees receiving each universe score is obtained by multiplying the total number of examinees in the population by the probability associated with each universe score. For example, if there were 1,000 students in the population, 24 students would be expected to spell correctly 20 words of the 50 words in the universe.

Table 3-13. Estimated Relative Frequency Per Universe Score on the 50-Item Universe Using the Negative Hypergeometric Distribution

Score	*Relative Frequency*	*Score*	*Relative Frequency*
0	.02305		
1	.02481	26	.02130
2	.02560	27	.02087
3	.02605	28	.02043
4	.02630	29	.01996
5	.02643	30	.01949
6	.02648	31	.01900
7	.02647	32	.01849
8	.02641	33	.01796
9	.02631	34	.01741
10	.02617	35	.01685
11	.02601	36	.01627
12	.02582	37	.01566
13	.02560	38	.01503
14	.02537	39	.01437
15	.02512	40	.01368
16	.02484	41	.01297
17	.02456	42	.01221
18	.02425	43	.01141
19	.02394	44	.01056
20	.02360	45	.00966
21	.02325	46	.00867
22	.02289	47	.00759
23	.02251	48	.00638
24	.02212	49	.00496
25	.02172	50	.00315

Estimating Percentiles

Once the relative frequency for each universe score has been calculated, estimated percentiles of the normative distribution is accomplished easily. The first step required is that of computing the cumulative relative frequency (crf) associated with each universe score. Using the relative frequencies given in Table 3-13, for example, crf(0) = .02305, crf(1) = .02305 + 02481, crf(2) = .02305 + .02481 + .02560, . . . , and crf(50) = 1.0. The percentile associated with a particular universe score $P(T_i)$ is equal to crf(T_i) multiplied by 100. The universe score associated with a particular percentile is that score whose crf when multiplied by 100 approximates most closely the desired percentile. This same procedure is followed regardless of what probability distribution has been used to approximate the normative distribution of universe scores.

The System of Pearson Distributions

One system of distributions for graduating normative distributions which merits consideration is the system devised by Pearson (1895, 1901) and popularized by Elderton (1938) and Elderton and Johnson (1969). There are 12 basic distributions in the Pearson system and each is obtained as a solution to the differential equation

$$\frac{dy}{dx} = -y \frac{x + c_1}{c_0 + c_1 x + c_2 x^2} \tag{3.24}$$

where the coefficients c_0, c_1 and c_2 are functions of the second through fourth central moments. Each of the 12 distributions is determined exactly by the first four moments. The calculations involved in each distribution are given in detail by Elderton and Johnson (1969) as well as suggested guidelines in deciding which distribution to use in approximating a specific normative distribution. Because the relevant equations have been neatly enumerated by Elderton and Johnson and, additionally, by Kendall (1952), they are not listed again here. All computations, however, required for each distribution have been incorporated into the computer program given in Appendix A. The forms of the Pearson distributions vary widely and include *U*-shaped, *J*-shaped, negatively-skewed, positively-skewed, symmetrical, normal, and exponentially-shaped curves.

The feasibility of using the Pearson distributions with parameters estimated through multiple matrix sampling to approximate normative distributions of universe scores has been systematically investigated only recently. Brandenburg and Forsyth (1973a, 1973b) examined the relative merits of using the Pearson Type I distribution (one of the 12 distributions) and the negative hypergeometric distribution to approximate the normative distributions associated with selected standardized tests. Using existing item scores on standardized tests, they examined in their first investigation the relative merits of these two distributions in approximating normative distributions when the *parameters* were used in the relevant equations; in their second investigation, *estimates of parameters* obtained through postmortem item-examinee sampling were used. In both investigations, they found that the Pearson Type I distribution was superior generally to the negative hypergeometric distribution. Although the Pearson Type I distribution is more complicated mathematically, it is computerized easily and in view of the findings by Brandenburg and Forsyth one can expect that the Pearson distributions will receive greater attention in the future. It should be noted that the Pearson distributions, in contrast to the negative hypergeometric distribution, may be used with all item-scoring procedures.

Additional Graduation Procedures

The negative hypergeometric distribution and the system of Pearson curves are but two procedures for graduating normative distributions. Other procedures which may be used are the linear-prediction model developed by Kleinke (1972), the empirical Bayes' estimation procedure described by Lord (1969), and the innovative but unlabeled procedure developed by Hill (1972). One direction towards which the research on multiple matrix sampling will be going in the future is the detailed examination of models which may be used in graduating normative distributions.

Chapter Four

Computer Simulation of Multiple Matrix Sampling

Described in this chapter is a simulation model for multiple matrix sampling and the computer program written by the author implementing the model. Although the simulation model has several areas of application, the reader uninterested in such matters can safely bypass this chapter without loss of continuity.

DEVELOPMENT OF MODEL

Although the basic statistical framework for multiple matrix sampling was given by Hooke (1956), Lord (1960) and Wilks (1962), that which presented the idea most forcibly to most researchers was a series of empirical investigations conducted by Lord (1962), Plumlee (1964), and Cook and Stufflebeam (1967). What these three investigations had in common was the general procedure used to test multiple matrix sampling—they all used postmortem item-examinee sampling. In this procedure, the scores obtained by testing all examinees are obtained in advance and the researcher samples from this data base acting as if only certain examinees had responded to certain items. Estimates of parameters obtained through multiple matrix sampling are then contrasted with the known parameters of the data base. The major problem with investigations like these is that all results are specific to the data base used by the investigator. Inherent in all conclusions drawn is the latent question, "What would have been the result if a data base having different values for parameters had been used?"

It goes almost without saying that conclusions about multiple matrix sampling obtained by algebraic or statistical derivation are preferred over those determined empirically. However, in the early development of many topics the variables of primary importance are not known and it is in identifying such variables that a simulation model has immediate application. Multiple matrix sampling is a developing topic and one whose pace of development is facilitated through simulation. Although the simulation model developed by the author and described in this chapter is essentially a postmortem sampling model, the technique is an improvement over previous postmortem investigations using "real" data because data bases having a wide variety of characteristics may be generated very easily. Consider the problem of investigating, for example, the effects of variation in the variance of item difficulty indices on the standard error of estimate for several parameters using "real" data. In this case, data bases would have to be located having the prescribed number of items, the prescribed reliability, the prescribed parameters, and having additionally the prescribed variations in the variance of the item difficulty indices. It is in the light of such a difficult task as this that the benefits of a simulation model are viewed most appropriately.

USES FOR COMPUTER SIMULATION PROGRAM

The computer program simulating multiple matrix sampling has application to several problem areas. Three noteworthy applications are: (a) approximating the standard error of estimate for any parameter of interest, (b) testing the feasibility of newly derived estimation procedures, and (c) providing students with a vehicle for learning about how multiple matrix sampling operates and what is to be expected in practice. Each of these three applications will be llustrated subsequently in showing how the model was used by Shoemaker (1972) in determining the feasibility of the jackknife procedure in approximating standard errors of estimate in multiple matrix sampling.

DESCRIPTION OF SIMULATION MODEL

The algorithm used within the model is described most appropriately as simulated postmortem item-examinee sampling. In postmortem sampling, item-examinee samples are taken from an N by K data base obtained by testing N examinees over K items; in *simulated* postmortem sampling, the N by K data base is computer-generated by a simulation model. As an overview, the computer program generates a data base (K item scores for each examinee), selects multiple item-examinee samples from this data base, performs all calculations necessary for estimating parameters, and replicates this procedure as many times as specified by the user before computing the statistics of the results obtained over replications. The computer program is restricted to data bases in which items are scored dichotomously and, in multiple matrix sampling, to subtests having an equal number of items administered to an equal number of examinees. Additionally, the model used assumes that the content within the universe of interest is homogeneous, that is, the model does not permit or incorporate any procedure for stratifying the universe on item content or item difficulty level or both. The model will undoubtedly be expanded in the future to permit such stratification procedures and will probably use the procedure for simulating content and difficulty stratification developed previously by Shoemaker and Osburn (1970).

Generation of Data Bases

In simulating multiple matrix sampling, generation of the data base is of primary importance. Although one procedure might be that of generating an N by K matrix and storing it in memory, a more appropriate procedure is one in which the item scores on the K-item universe are generated for one and only one examinee at a time. All that is stored in memory are the K item scores for *one* examinee. The procedure, however, for generating item scores must be one such that over any number of hypothetical examinees generated the items and universe scores have prescribed characteristics. In this procedure, the population of examinees is taken to be infinite. The parameters subject to manipulation within the program are: (a) K, the number of items in the universe, (b) μ'_1, the mean universe score, (c) μ_2, the variance of universe scores, (d) KR_{20}, the coefficient of reliability for the universe scores, (e) σ^2_I, the variance of the item difficulty indices where the difficulty index for an item is the proportion of examinees answering the item correctly, and (f) the degree of skewness in the distribution of universe scores.

These six parameters of the data base are not independent and are restricted both by the limitations of the computer program and inherent statistical relationships. In the computer program, values for K, μ'_1 and σ^2_I must be specified by the user. Because the maximum permitted value for K is 150 (the program, however, may be modified to incorporate larger values of K), μ'_1 is restricted to values between 0 and K. If KR_{20} is specified by the user, μ_2 is determined by the well-known relationship

$$KR_{20} = \frac{K}{K-1}\left[\frac{\mu_2 - \mu'_1 + \mu'^2_1/K + K\sigma^2_I}{\mu_2}\right] \qquad (4.1)$$

derived originally by Tucker (1949). If μ_2 is specified by the user, KR_{20} is determined consequently. Such an arrangement has been incorporated within the program to facilitate hypothesis testing where either μ_2 or KR_{20} is to be controlled across levels of K. Of course, $\bar{p} = \mu'_1/K$ is determined once μ'_1 has been specified.

The degree of skewness in the normative distribution is simulated by using the lognormal or normal probability distribution functions to generate test score distributions. The lognormal distribution with two parameters is used to generate positively-skewed test score distributions while the three-parameter lognormal distribution is used for negatively-skewed distributions. The lognormal distribution is described in detail by Aitchison and Brown (1957) and a detailed explanation of simulating stochastic variates with the lognormal distribution is given by Naylor, Balintfy, Burdick, and Chu (1966). The normal density function is, of course, used to simulate normal test score distributions. Density functions for the two and three parameter lognormal probability distributions are, respectively,

$$LN(T \mid \mu, \sigma^2) = \frac{1}{T\,\sigma(2\,\pi)^{1/2}} \exp\left[-\frac{\{\ln(T) - \mu\}^2}{2\sigma^2}\right] \tag{4.2}$$

$$LN(T' \mid T' = K - T, \mu, \sigma^2) = \frac{1}{T'\,\sigma(2\pi)^{1/2}} \exp\left[-\frac{\{\ln(T') - \mu\}^2}{2\sigma^2}\right] \tag{4.3}$$

For the normal distribution, the density function is

$$ND(T \mid \mu, \sigma^2) = \frac{1}{\sigma(2\,\pi)^{1/2}} \exp\left[-\frac{(T - \mu)^2}{2\sigma^2}\right] \tag{4.4}$$

The constants μ and σ^2 in Equation (4.4) are equal, respectively, to the desired mean and variance in the normative distribution; however, in Equations (4.2) and (4.3), μ and σ^2 are a *function of* the desired mean and variance of universe scores. Specifically, the μ and σ^2 in Equations (4.2) and (4.3) are computed by

$$\mu = \ln(\mu_1') - \frac{\ln(\mu_2/\mu_1'^2 + 1)}{2} \tag{4.5}$$

and

$$\sigma^2 = \ln(\mu_2/\mu_1'^2 + 1). \tag{4.6}$$

The appropriate derivations for (4.5) and (4.6) are given by Naylor, Balintfy, Burdick and Chu (1966). If z is a random normal deviate ND(0,1), universe scores T having lognormal distributions are generated by

$$T_i = \exp(\mu + \sigma z_i) \qquad \text{for } i = 1, 2, \ldots, N \tag{4.7}$$

for positively-skewed distributions, and

$$T_i = K - \exp(\mu + \sigma z_i) \qquad \text{for } i = 1, 2, \ldots, N \tag{4.8}$$

for negatively-skewed distributions. For normal distributions,

$$T_i = \mu + \sigma z_i \qquad \text{for } i = 1, 2, \ldots, N. \tag{4.9}$$

The T_i computed in (4.7), (4.8), and (4.9) will be continuous variables. Because items are scored dichotomously, the T score must be rounded to the nearest integer value. The midpoint of each score interval is taken to be that point above which one-half of the area in that score interval is found. This point is found by integrating via trapezoid rule the area under the appropriate normal or lognormal curve. If the T score is equal to or greater than the midpoint, the score is rounded up; if not, the score is rounded down.

Item scores are related to test scores. Specifically, if X_{ij} is the item score for examinee i on item j, $\sum_j^K X_{ij} = T_i$. Also, $\overline{p} = \overline{T}/K$. If σ_I^2 is greater than zero, individual item difficulty indices are generated by

$$p_i = \overline{p} + \sigma_I z_i \qquad \text{for } i = 1, 2, \ldots, K \tag{4.10}$$

where z_i is a random normal deviate. When σ_I^2 is not equal to zero, the distribution of p_i values will be approximately normal. If σ_I^2 is equal to zero, $p_i = \overline{p}$ for all values of i. With markedly skewed distributions, σ_I^2 is typically $0 < \sigma_I^2 < .001$ and, because of this, σ_I^2 is set to zero for all skewed distributions generated by the simulation model. After the item difficulty indices have been generated within the program, deciding if an examinee passes or fails each item is relatively simple. Item difficulty indices are computed for all items generated. An examinee "passes" those items which will bring the computed item difficulty indices most closely to the desired item difficulty indices. For example, if the computed item difficulty for item i were less than the desired item difficulty for item i, examinee j would pass item i; if the computed difficulty were equal to or greater than the desired item difficulty, he "fails" item i. In the program, the desired item difficulty indices are sorted in descending order. If, in following the algorithm from the first through the Kth item, $\sum^K X_{ij} \neq T_j$, the first $T_j - \sum^K X_{ij} = d$ items not already passed by examinee j are scored by the program as items answered correctly by him.

Validity of Simulation Model

The validity of the simulation model is found in its ability to generate the desired data base. Two examples of data bases generated by the model are given in Tables 4-1 and 4-2. Although the discrepancies in Table 4-2 are minor, it should be noted that the magnitude of the discrepancies decreases with increases in K.

Table 4-1. Results Obtained From Simulation Model For 3000 Examinees When K = 20 With the Normative Distribution Distributed Normally

Parameter	*Input*	*Output*
K	20	20
μ_1'	10.0000	10.0150
μ_2		computed
KR_{20}	.8000	.7999
$\overline{p}$	.5000	.5010
σ_I^2	.0800	.0799
N	3000	3000

Item	*Item Difficulty Indices* *Input*	*Output*	*Obtained Frequency Dist.* *Score*	*Frequency*
			0	16
1	.987	.987	1	17
2	.901	.901	2	36
3	.857	.857	3	47
4	.782	.787	4	96
5	.750	.750	5	136
6	.737	.738	6	182
7	.702	.703	7	225
8	.634	.634	8	307
9	.565	.566	9	266
10	.518	.519	10	328
11	.458	.458	11	327
12	.442	.442	12	256
13	.430	.430	13	210
14	.374	.374	14	191
15	.275	.275	15	122
16	.208	.209	16	104
17	.139	.140	17	67
18	.113	.114	18	38
19	.112	.113	19	16
20	.026	.026	20	13

Simulation of Multiple Matrix Sampling

Subtests are constructed within the program by sampling at random k items from the K-item universe and two procedures—described more appropriately as restricted random sampling—may be used. In the first, items are sampled randomly from the universe without replacement within each subtest but with replacement among subtests; in the second, without replacement within each subtest and without replacement among subtests. The major difference between these two strategies is that, in the first, an individual item may appear in more than one subtest whereas this is not possible with the second strategy. After the

Table 4-2. Results Obtained From Simulation Model For 3000 Examinees When K = 20 With the Normative Distribution Negatively-Skewed (Three Parameter Lognormal Distribution)

Parameter	*Input*	*Output*
K	20	20
μ_1'	17.5000	17.6150
μ_2		computed
KR_{20}	.8000	.7570
$\overline{p}$	.8750	.8810
σ^2_I	.0000	.0002
N	3000	3000

Item	*Item Difficulty Indices* *Input*	*Output*
1	.875	.930
2	.875	.908
3	.875	.888
4	.875	.876
5	.875	.876
6	.875	.876
7	.875	.876
8	.875	.876
9	.875	.876
10	.875	.876
11	.875	.876
12	.875	.876
13	.875	.876
14	.875	.876
15	.875	.876
16	.875	.876
17	.875	.876
18	.875	.876
19	.875	.875
20	.875	.875

Obtained Frequency Dist. *Score*	*Frequency*
0	16
1	3
2	2
3	5
4	4
5	3
6	3
7	6
8	7
9	8
10	15
11	32
12	32
13	43
14	84
15	135
16	200
17	352
18	670
19	885
20	495

items have been allocated to subtests, the item scores on the K-item universe are generated sequentially for n examinees. Only those k items are considered which were selected for that subtest. This process is repeated for each subtest with the computations performed on each item-examinee sample being identical to those given in Chapter 3. If the user specifies r replications of a particular sampling plan ($t/k/n$), r pooled estimates of each parameter will be produced with the standard error of estimate per parameter associated with that sampling plan calculated as the standard deviation of the r pooled estimates of each parameter.

COMPUTING PARAMETERS OF DATA BASE

In using the model, the user must specify K, σ_I^2, μ_1' and either KR_{20} or μ_2. If KR_{20} is specified, μ_2 is computed as

$$\mu_2 = \frac{\mu_1'(K - \mu_1') - K^2 \sigma_I^2}{K - (K-1)KR_{20}} \tag{4.11}$$

Noting that $\mu_2 = K\sigma_{IE}^2 + K^2 \sigma_E^2$, the components of variance of the normative data base are calculated by

$$\sigma_E^2 = \frac{KR_{20}\mu_2}{K^2} \tag{4.12}$$

and

$$\sigma_{IE}^2 = \frac{\mu_2(1 - KR_{20})}{K}. \tag{4.13}$$

The reliability of the average item r_{11} is equal to

$$r_{11} = \frac{\theta'}{1 + \theta'} \tag{4.14}$$

where

$$\theta' = \frac{KR_{20}}{K(1 - KR_{20})}.$$

The third and fourth central moments for normal normative distributions are equal to

$$\mu_3 = 0 \tag{4.15}$$

and

$$\mu_4 = 3\mu_2^2. \tag{4.16}$$

For lognormal distributions, these moments are equal to

$$\mu_3 = \mu_1'^3 \left[(e^z - 1)^3 + 3(e^z - 1)^2 \right] \tag{4.17}$$

and

$$\mu_4 = \mu_1'^4 \left[(e^z - 1)^6 + 6(e^z - 1)^5 + 15(e^z - 1)^4 + 16(e^z - 1)^3 + 3(e^z - 1)^2 \right] \tag{4.18}$$

where

$$z = \ln \left(\frac{\sigma^2}{\mu^2} + 1 \right)$$

If the lognormal distribution is skewed negatively, $\mu_3 = -\mu_3$.

In the simulation model used here, knowing the values of the parameters of the data base makes it possible to more easily evaluate the estimates obtained from any particular sampling plan. When these values are included in a table (as, for example, Tables 4-3 and 4-7) they will be labeled as a "Norm" value.

SAMPLE OUTPUT FROM SIMULATION PROGRAM

The potential of the simulation program is illustrated well through considering its use by Shoemaker (1972) in determining the feasibilty of the jackknife procedure (described in Chapter 3) in approximating standard errors of estimate in multiple matrix sampling. The parameters of the data base manipulated systematically in this investigation were: (a) the number of test items (K = 40, 60), (b) variance of the item difficulty indices (σ_I^2 = .00, .05), and (c) degree of skewness in the normative distribution (distributed normally, markedly negatively skewed). Nine item-examinee sampling plans were used on each data base. The reliability of the total scores was equal to .80 for all data bases generated. The procedure used to test the jackknife was relatively straightforward. Because each sampling plan was replicated r (50) times, r estimates of each parameter were produced as well as r estimates of the jackknifed standard error for each parameter. At the end of r replications, *two* estimates of the standard error of estimate for each parameter for each sampling plan were computed. The first estimate was obtained by computing the standard deviation of the r pooled estimates of each parameter; the second, by computing the mean of the r jackknifed standard errors for each parameter. The jackknife is considered appropriate then to the degree that the two standard errors agree.

The results for one sampling plan using a normal normative distribution of test scores with the number of items in the universe equal to 40 and the variance of the item difficulty indices equal to .05 are reported here. The mean test score was set at 20 and the reliability of the 40-item universe was equal to .80 ($\theta' = .10$). The specific sampling plan considered here as an example was one requiring four subtests having ten items each with each subtest administered to 50 examinees. This particular sampling plan (like the other nine) was replicated 50 times. Items for each subtest within each replication were selected without replacement for each subtest and without replacement among subtests. Results obtained from the first two replications are given in Tables 4-3 and 4-4 and the pooled results over five replications and 50 replications are given, respectively, in Tables 4-5 and 4-6. Results obtained over 50 replications with the same sampling plan but sampling items among subtests *with replacement* (instead of without replacement) are given in Table 4-7.

Although the results reported in Tables 4-3 through 4-7 are only a

Table 4-3. Results Obtained From First Replication of Sampling Plan (Items Sampled Without Replacement)

	*Estimates Of Moments**				*Components Of Variance*			
Subtest	$\hat{\mu}'_1$	$\hat{\mu}_2$	$\hat{\mu}_3$	$\hat{\mu}_4$	$\hat{\sigma}^2_I$	$\hat{\sigma}^2_E$	$\hat{\sigma}^2_{IE}$	$\hat{\theta}'$
1	21.840	19.626	90.168	1467.4	.061	.008	.186	.042
2	19.440	22.511	-121.640	1795.4	.047	.009	.199	.046
3	29.120	22.657	-139.480	238.5	.077	.009	.171	.059
4	14.880	66.651	46.850	9356.3	.032	.037	.169	.224
Pooled Estimate	19.320	32.861	-31.025	3214.4	.054	.016	.181	.093
SE	1.563	11.285	58.257	2074.5	.001	.007	.007	.044
Norm	20.000	36.364	0.000	3966.938	.050	.018	.200	.100

*Estimates of moments were obtained using equations given by Lord (1960)

Table 4-4. Results Obtained From Second Replication of Sampling Plan (Items Sampled Without Replacement)

	*Estimates Of Moments**				*Components Of Variance*			
Subtest	$\hat{\mu}'_1$	$\hat{\mu}_2$	$\hat{\mu}_3$	$\hat{\mu}_4$	$\hat{\sigma}^2_I$	$\hat{\sigma}^2_E$	$\hat{\sigma}^2_{IE}$	$\hat{\theta}'$
1	18.720	24.028	130.370	3179.0	.068	.011	.178	.060
2	21.680	33.702	-263.570	3303.8	.050	.016	.188	.088
3	20.960	31.471	75.877	4880.0	.048	.015	.192	.078
4	19.760	67.180	-91.493	9444.7	.023	.037	.193	.193
Pooled Estimate	20.280	39.095	-37.205	5201.9	.047	.020	.188	.105
SE	.654	9.587	89.003	1466.3	.009	.006	.004	.030
Norm	20.000	36.364	0.000	3966.938	.050	.018	.200	.100

*Estimates of moments were obtained using equations given by Lord (1960)

Table 4-5. Pooled Results Obtained Over Five Replications of Sampling Plan (Items Sampled Without Replacement)

	Parameter (Norm)	*Mean Estimate*	*SE (Replications)*	*Mean SE (Jackknife)*	*SD (SE(Jackknife))*
μ'_1	(20.000)	19.748	.398	1.180	.336
μ_2	(36.364)	34.396	2.614	8.042	2.664
μ_3	(0.000)	−9.865	39.628	92.923	28.768
μ_4	(3966.938)	4149.996	703.058	1812.603	214.401
σ^2_I	(.050)	.052	.007	.011	.004
θ'	(.100)	.095	.008	.028	.011
KR_{20}	(.800)	.791	.014	–	–

Table 4-6. Pooled Results Obtained Over Fifty Replications of Sampling Plan (Items Sampled Without Replacement)

	Parameter (Norm)	*Mean Estimate*	*SE (Replications)*	*Mean SE (Jackknife)*	*SD (SE(Jackknife))*
μ'_1	(20.000)	19.990	.403	1.462	.496
μ_2	(36.364)	37.237	7.149	7.806	3.107
μ_3	(0.000)	−.709	81.021	90.183	39.032
μ_4	(3966.938)	4022.215	1653.719	1613.260	875.725
σ^2_I	(.050)	.050	.006	.011	.005
θ'	(.100)	.108	.030	.030	.015
KR_{20}	(.800)	.804	.038	–	–

Table 4-7. Pooled Results Obtained Over Fifty Replications of Sampling Plan (Items Sampled With Replacement)

	Parameter (Norm)	*Mean Estimate*	*SE (Replications*	*Mean SE (Jackknife)*	*SD (SE(Jackknife))*
μ'_1	(20.000)	19.999	1.548	1.280	.602
μ_2	(36.364)	36.595	7.954	8.908	3.497
μ_3	(0.000)	7.668	92.411	82.287	56.552
μ_4	(3966.938)	4237.750	1883.277	1820.540	1050.421
σ^2_I	(.050)	.051	.011	.010	.004
θ'	(.100)	.108	.033	.037	.019
KR_{20}	(.800)	.800	.054	–	–

small part of the composite collected by Shoemaker, they nevertheless illustrate both results obtained from the simulation model and how the results may be used to gain insights into multiple matrix sampling. The results in Tables 4-3 and 4-4 illustrate the variability of estimates in going from one replication to another. Contrasting the results in Table 4-5 with those given in Table 4-6 suggests that estimates of standard errors are still relatively unstable at five replications and that, as would be expected, a more accurate assessment of standard errors is possible with 50 replications.

The results in Tables 4-6 and 4-7 suggest that on the average the jackknife approximates well the standard error of estimate. The one exception to this is found in estimating the standard error of the mean universe score when items are sampled without replacement among subtests. Here, the jackknife overestimates the standard error of estimate. This result is not surprising because the jackknife assumes random sampling with replacement among subtests; however, this violation of assumption does not appear to present a problem when estimating standard errors of higher moments. The standard error of estimate for the mean universe score was computed using Equation (3.15a) for each of the 50 replications of the sampling plan when items were sampled without replacement. (The equation is not applicable when items are sampled with replacement among subtests.) The mean of these 50 estimates was .542 and their standard deviation (or standard error of estimate for that equation) was .031. The .403 reported in Table 4-6 is in line with this result. However, the standard error of the jackknife procedure in estimating the standard error of the mean universe score was .496 (disregarding its bias on the average here) and, in contrasting this with the .031 standard error of Equation (3.15a) suggests that one of the hazards of the jackknife may be its variability of results.

The results presented in Tables 4-3 and 4-7 were given to illustrate how the simulation program may be used to gain insights into multiple matrix sampling. In this case, the program was used to determine the feasibility of the jackknife procedure in approximating standard errors of estimate. This is only one of the many uses which may be made of the program. A detailed listing of the computer program is given in Appendix A.

Chapter Five

Hypothesis Testing

Various hypotheses may be tested through multiple matrix sampling and the emphasis of this chapter is on how the procedure may be used for this purpose. The intent of the chapter is not to provide an exhaustive listing of all hypotheses which may be tested through multiple matrix sampling but is, instead, an examination of several hypotheses most likely to be tested in any investigation using the procedure. Although hypotheses regarding means, central moments, and distributions are considered here, this by no means is meant to imply that these are the only hypotheses to which multiple matrix sampling is applicable. The opposite is certainly the case and undoubtedly one of the major thrusts of future research in multiple matrix sampling will be an expansion of the procedure into a wide variety of hypothesis testing techniques.

HYPOTHESES REGARDING MEAN UNIVERSE SCORES

After having estimated the mean test score over an item universe for one or more examinee populations, the investigator is interested frequently in determining if, for the case of one population, the estimated mean differs significantly from some value or, for the case of several populations, if any or all of the estimated means differ significantly among themselves. An example of the former problem is an investigation concerned with determining if the mean performance of students within a school district differs significantly from the standard (test score) defining the minimally acceptable performance. The latter problem is found in an investigation focused on determining the differences among three competing reading programs used by a particular school district at a particular grade level. It goes almost without saying that, in the first example, the estimated mean will not be exactly equal to the standard and, in the second example, that the three estimated means will not be exactly equal. In both examples, there will be discrepancies and it is the task of the researcher to determine the smallest observed discrepancy which, in his opinion, is indicative of a true difference between a mean and a standard or among several means. Sirotnik (1973) has considered this problem in some detail and the discussion in this section reflects generally his analysis.

Hypotheses Regarding One Mean
$(H_0: \mu'_1 = c)$
In testing the hypothesis that $\mu'_1 = c$, the statistic w

$$w = \frac{\hat{\mu}'_{1p} - c}{SE(\hat{\mu}'_{1p})} \tag{5.1}$$

may be computed in three ways with the difference among the procedures due to the manner in which the standard error of estimate for the pooled estimate of the mean universe score is computed. If the items for each subtest are sampled with replacement among subtests, the standard error is computed most appropriately by taking the square root of

$$VAR(\hat{\mu}'_{1p}) = \frac{Q_1}{tkn}\ , \tag{5.2}$$

where,

$Q_1 = \mathrm{MS}_{E(B)} + \mathrm{MS}_{I(B)} - \mathrm{MS}_{I(B)E(B)}$
t = number of subtests,
k = number of items per subtest, and
n = number of examinees per subtest.

The w defined by (5.1) is distributed according to the t-distribution with degrees of freedom equal to the nearer integer value of

$$\frac{Q_1^2}{\dfrac{MS^2_{E(B)}}{t(n-1)} + \dfrac{MS^2_{I(B)}}{t(k-1)} + \dfrac{MS^2_{E(B)I(B)}}{t(n-1)(k-1)}}$$

(It should be noted here that when items are scored dichotomously, the w defined by (5.1) is distributed only approximately according to the t-distribution. This point is discussed in greater detail on page 74.)

The rationale behind (5.2) is explained readily by casting the problem into the framework of the analysis of variance (AOV). Within this framework, the total variation among item scores is attributable to five sources, as indicated in Table 5-1 where B refers to subtests, E to examinees, and I to items. The notation $E(B)$ indicates that examinees are nested within subtests; $I(B)$

Table 5-1. Analysis of Variance Sources for One Population When Sampling Items With Replacement Among Subtests

Source	*df*	*EMS*
Mean	1	$\sigma^2_{EI} + k\sigma^2_E + n\sigma^2_I + tkn\mu^2$
B	(t-1)	$\sigma^2_{EI} + k\sigma^2_E + n\sigma^2_I + nk\sigma^2_B$
E(B)	t(n-1)	$\sigma^2_{EI} + k\sigma^2_E$
I(B)	t(k-1)	$\sigma^2_{EI} + n\sigma^2_I$
E(B)I(B)	t(n-1)(k-1)	σ^2_{EI}

indicates similarly that items are nested within subtests. In this problem (and in multiple matrix sampling generally), items and examinees are considered "nested" within subtests instead of "crossed" with subtests because different examinees respond to different subtests. Although the component of "variance" associated with the mean universe score is equal to $(\sigma^2_{EI} + k\sigma^2_E + n\sigma^2_I)/tkn$, no source listed in Table 5-1 has an expected mean square (EMS) equal exactly to the numerator of this fraction. This numerator, however, may be computed by adding together three sources of variation–*E(B)*, *I(B)*, and *E(B)I(B)*. The expected mean square of *E(B)* + *I(B)* + *E(B)I(B)* is equal to $\sigma^2_{EI} + k\sigma^2_E + \underline{\sigma^2_{EI}} + n\sigma^2_I - \underline{\sigma^2_{EI}}$. The underscored components of variance cancel out of the equation leaving the desired result. This procedure for forming expected mean squares and determining the associated degrees of freedom was described originally by Satterthwaite (1946).

Equation (5.2) is applicable to all item-scoring procedures. A problem exists, however, in that (5.2) is strictly appropriate only when items have been sampled with replacement among subtests–a procedure used rarely in practice. (In multiple matrix sampling, items are sampled most commonly without replacement among subtests.) When items are sampled without replacement among subtests, the general linear model upon which the EMS given in Table 5-1 were derived is no longer applicable and Equation (5.2) will overestimate the standard error of the mean. The degree of overestimation is a function of the magnitude of *K* and *N* and to date has not been determined either theoretically or empirically.

The second procedure for computing the standard error of the pooled estimate of the mean universe score is defined by Equations (3.14), (3.15) or (3.15a) given in Chapter 3. These equations were derived on the assumption that items for each subtest are sampled randomly and without replacement. When this is the case, these equations are preferred over (5.2) in estimating the standard error of the pooled estimate of the mean universe score and the *w* defined by Equation (5.1) is distributed normally ND(0,1).

Thirdly, the jackknife procedure may be used to compute the standard error of the pooled estimate of the mean universe score. The jackknife procedure, described in detail in Chapter 3, is strictly appropriate when items are sampled with replacement among subtests and gives an overestimate of the standard error when items are sampled without replacement (Shoemaker, 1973). When the jackknife is used, the w defined by Equation (5.1) is distributed approximately according to the t distribution with $(t - 1)$ degrees of freedom. For the jackknife procedure to be appropriate here, a large number of subtests should be used.

Hypotheses Regarding Several Means ($H_O: \mu'_{1\,1} = \mu'_{1\,2} = \ldots \mu$)

The analysis of variance procedure may be used appropriately to test the hypothesis of equality among means for two or more populations. Assuming that the number of populations selected for investigation constitutes those of interest (and not a random sample from those of interest), the sources of variation and expected mean squares for this design are listed in Table 5-2. The EMS are those given by Sirotnik (1973) using the procedure devised originally by Cornfield and Tukey (1956). In Table 5-2, A refers to the number of populations being contrasted and E, I, and B have the same meaning as before.

Table 5-2. Analysis Of Variance Sources For Several Populations When Items Are Sampled With Replacement Among Subtests

Source	*df*	*EMS*
A	(a – 1)	$\sigma^2_{EI} + k\sigma^2_E + n\sigma^2_{AI} + tkn\sigma^2_A$
B	(t – 1)	$\sigma^2_{EI} + k\sigma^2_E + an\sigma^2_I + ank\sigma^2_B$
E(AB)	at(n – 1)	$\sigma^2_{EI} + k\sigma^2_E$
I(B)	t(k – 1)	$\sigma^2_{EI} + an\sigma^2_I$
AB	(a – 1)(t – 1)	$\sigma^2_{EI} + k\sigma^2_E + n\sigma^2_{AI} + nk\sigma^2_{AB}$
AI(B)	t(a – 1)(k – 1)	$\sigma^2_{EI} + n\sigma^2_{AI}$
E(AB)E(B)	at(n – 1)(k – 1)	σ^2_{EI}

In the analysis of variance, the F-ratio for testing the differences among the A levels is constructed by dividing the calculated mean square for A (MS_A) by a source of variation whose expected mean square is equal to $\sigma^2_{EI} + k\sigma^2_E + n\sigma^2_{AI}$. Because no single source of variation has this expected mean square, one must be pieced together using the technique described by Satterthwaite (1946). In this case,

$$Q_2 = MS_{E(AB)} + MS_{AI(B)} - MS_{E(AB)I(B)}$$

and the F-ratio (now referred to as a quasi F-ratio because the denominator was pieced together) is equal to

$$F' = \frac{MS_A}{Q_2} . \tag{5.3}$$

The degrees of freedom associated with the denominator in (5.3) are equal to the nearer integer value of

$$\frac{Q_2^2}{\dfrac{MS^2_{E(AB)}}{at(n-1)} + \dfrac{MS^2_{AI(B)}}{t(a-1)(k-1)} + \dfrac{MS^2_{E(AB)I(B)}}{at(n-1)(k-1)}}$$

To test B, the quasi F-ratio is formed by dividing MS_B by Q_3, where

$$Q_3 = MS_{E(AB)} + MS_{I(B)} - MS_{E(AB)I(B)} .$$

The degrees of freedom associated with the denominator of this quasi F-ratio are equal to the nearer integer value of

$$\frac{Q_3^2}{\dfrac{MS^2_{E(AB)}}{at(n-1)} + \dfrac{MS^2_{I(B)}}{t(k-1)} + \dfrac{MS^2_{E(AB)I(B)}}{at(n-1)(k-1)}}$$

Similarly, to test the A by B interaction, the quasi F-ratio is formed by dividing MS_{AB} by Q_4, where

$$Q_4 = MS_{AI(B)} + MS_{E(AB)} - MS_{E(AB)I(B)} .$$

The degrees of freedom associated with Q_4 are equal to the nearer integer value of

$$\frac{Q_4^2}{\dfrac{MS^2_{AI(B)}}{t(a-1)(k-1)} + \dfrac{MS^2_{E(AB)}}{at(n-1)} + \dfrac{MS^2_{E(AB)I(B)}}{at(n-1)(k-1)}}$$

A good description of quasi F-ratios and procedures for deriving them are given by Kirk (1968).

An example may be helpful at this point in illustrating the computations involved in the quasi F-ratios and in the calculation of the mean squares for each source of variation. The statistical layout of a hypothetical problem is given in Figure 5-1 and its associated data (item scores) are given in Table 5-3. The item-examinee sampling plan involves three subtests containing two items each administered to five examinees selected randomly from the population. In Table 5-5 are given the computed sums of squares (SS) and mean squares (MS). It should be noted that, if an equal number of examinees responds to each subtest, the nested SS may be computed easily by any of the standard AOV programs available in most computer program libraries, i.e., BMD02V or BMD05V (Dixon, 1968). The trick here is to act as if the problem were a four-factor AOV design with all factors crossed. The first three factors are A (treatment groups), B (subtests), and I (items); the fourth factor is E (examinees). The resultant SS are then

		A_1						A_2						A_3			
B_1		B_2		B_3		B_1		B_2		B_3		B_1		B_2		B_3	
I_1	I_2	I_3	I_4	I_5	I_6	I_1	I_2	I_3	I_4	I_5	I_6	I_1	I_2	I_3	I_4	I_5	I_6
E_1	E_1	E_6	E_6	E_{11}	E_{11}	E_{16}	E_{16}	E_{21}	E_{21}	E_{26}	E_{26}	E_{31}	E_{31}	E_{36}	E_{36}	E_{41}	E_{41}
E_2	E_2	•	•	•	•	•	•	•	•	•	•	•	•	•	•	•	•
		•	•	•	•	•	•	•	•	•	•	•	•	•	•	•	•
E_3	E_3	•	•	•	•	•	•	∘	•	•	•	•	•	•	•	•	•
E_4	E_4																
E_5	E_5	E_{10}	E_{10}	E_{15}	E_{15}	E_{20}	E_{20}	E_{25}	E_{25}	E_{30}	E_{30}	E_{35}	E_{35}	E_{40}	E_{40}	E_{45}	E_{45}

Figure 5-1. Statistical Layout for Analysis of Variance Problem Involving Three Treatment Groups (A)

Table 5-3. Hypothetical Data for Sample Problem

		A_1						A_2						A_3			
B_1		B_2		B_3		B_1		B_2		B_3		B_1		B_2		B_3	
I_1	I_2	I_3	I_4	I_5	I_6	I_1	I_2	I_3	I_4	I_5	I_6	I_1	I_2	I_3	I_4	I_5	I_6
0	0	0	0	0	0	0	1	0	0	0	0	0	0	1	1	0	0
0	1	0	0	0	0	0	0	1	1	0	0	1	0	0	1	1	0
0	0	1	1	0	0	1	0	0	1	1	0	1	0	1	0	1	1
1	0	0	1	1	0	1	0	1	0	1	1	1	1	0	1	1	1
1	0	1	0	1	1	1	1	0	1	1	1	1	1	1	1	1	1

pooled as indicated in Table 5-4 to obtain the nested SS. For example, $SS_{E(AB)} = SS_E + SS_{AE} + SS_{BE} + SS_{ABE}$. These and some additional relationships between nested sums of squares and sums of squares when factors are crossed are given in detail by Winer (1971) and make what is normally a laborious computation quite easy. Also, no new computer programming is required. The computed SS and MS for the problem in hand are given in Table 5-5.

In testing the differences among the A treatments, the quasi F-ratio (denoted by F') is equal to

$$F' = \frac{MS_A}{Q_2} = \frac{.83333}{.10000} = 8.33330,$$

Table 5-4. Relationship Between Nested SS and Crossed SS in Sample Problem

Source	*E(AB)*	*I(B)*	*AI(B)*	*E(AB)I(B)*
A				
B				
I		+		
E	+			
AB				
AI			+	
AE	+			
BI		+		
BE	+			
IE				+
ABI			+	
ABE	+			
AIE				+
BIE				+
ABIE				+

Table 5-5. Analysis of Variance for Sample Problems

Source	*df*	*SS*	*MS*
A	2	1.66667	.83333
B	2	.06667	.03333
E(AB)	36	11.20000	.31111
I(B)	3	.96667	.32222
AB	4	.06667	.01667
AI(B)	6	.13333	.02222
E(AB)I(B)	36	8.40004	.23333
Total	89	22.50000	

where,

$$Q_2 = MS_{E(AB)} + MS_{AI(B)} - MS_{E(AB)I(B)}$$
$$= .31111 + .02222 - .23333$$
$$= .10000.$$

The degrees of freedom associated with Q_2 are equal to

$$df_{Q_2} = \frac{.10000^2}{\frac{.31111^2}{36} + \frac{.02222^2}{6} + \frac{.23333^2}{36}}$$
$$= 2.33481 \text{ or } 2.$$

The expected mean squares given in Table 5-2 are valid only if items are sampled with replacement among subtests or if the number of items in the universe is quite large. If items per subtest are sampled without replacement, the EMS given in Table 5-2 overestimate the error variance resulting in a conservative estimate of the F'-ratios. At first glance, it may seem that the difficulty involved in sampling items without replacement can be avoided completely by using subtest scores in place of item scores. This is not the case and doing so is exactly equal to using $MS_{E(AB)}$ as the error term for MS_A in Table 5-5.

Analysis Of Variance With Dichotomous Scores

The analysis of variance procedures described in this section for hypothesis testing are applicable for all item scoring procedures—dichotomous and polychotomous. However, when items are scored dichotomously, a complication arises of which the researcher must be aware. Specifically, a dichotomous variable (one which may be equal to one or the other of two values) is a violation of the assumptions on which the F-ratio in the analysis of variance is based. Of chief concern is that the scores on a dichotomous variable are not distributed normally in the population or populations of interest. Empirical results reported by Lunney (1970) and Light and Margolin (1971) and analytical discussions of the problem by D'Agostino (1971) and Snedecor and Cochran (1967) suggest that the F-ratio is robust with regard to this violation of assumptions providing certain guidelines are followed:

1. The frequencies per cell must be equal.
2. The usual AOV procedure may be used if the proportions (e.g., the proportions of 1's and 0's) per cell range between .25 and .75 and the degrees of freedom for the error mean squares is at least 20.

3. If the proportions per cell are between .00 and .30 or between .70 and 1.00, then the item scores should be transformed before performing the analysis of variance. Two recommended transformations are the arcsin and the logit. Discussions of various transformations are given by D'Agostino (1971), Winsor (1948), Dyke and Patterson (1952), Gart and Zweifel (1967), and Cochran (1950).
4. If the proportions per category per cell are less than .30 (or greater than .70) but do not have a wide range, then any transformation of the data will not prove useful. In this case, the degrees of freedom for the error mean square should be at least 40.

HYPOTHESES REGARDING CENTRAL MOMENTS

Hypotheses involving the variance of universe scores may be tested through confidence intervals constructed using the pooled estimate of the variance, the standard error of estimate obtained from Equation (3.17), and the confidence interval defined by inequality (3.21). To demonstrate the procedure, consider the example in which student achievement on an item universe was estimated through multiple matrix sampling both at the start of an instructional program (Pre) and at the end (Post). The estimated variance of universe scores ($\hat{\mu}_{2_p}$) at pretest was computed to be 2242.2 with the .95 confidence interval ranging from 754.4 to 3730.1. At posttest, the estimated variance was equal to 1781.7 with the .95 confidence interval ranging from 467.3 to 3087.1. Because the two confidence intervals overlap, the variances are judged not to differ significantly at the .05 level. Computing confidence intervals for the third and fourth central moments of universe scores is more complicated and is discussed in Chapter 3 in the section entitled "Computing Confidence Intervals For Parameters."

HYPOTHESES REGARDING DISTRIBUTIONS

The normative distribution estimated through multiple matrix sampling provides the basic data for several goodness-of-fit tests. For example, the Kolmogorov-Smirnov one-sample test (Siegel, 1956) provides a test of the hypothesis that the approximated distribution of scores came from a population of scores having a specified theoretical distribution. The test involves specifying the cumulative frequency distribution which would occur under the theoretical distribution and comparing that with the approximated cumulative frequency distribution. A simple extension of the Kolmogorov-Smirnov one-sample test is the Kolmogorov-Smirnov two-sample test (Siegel, 1956) which is concerned with the agreement between two approximated frequency distributions.

Chapter Six

Specific Applications

The range of applications of multiple matrix sampling to research problems is, indeed, broad especially when the problem is cast into the "object-variable" framework described in Chapter 1. Although the examples cited in this chapter demonstrating how the procedure may be or has been used in practice are taken generally from the area of educational research, the implication should not be made that multiple matrix sampling is restricted solely to this domain. The choice of examples, however, does imply correctly that currently the area to which multiple matrix sampling is applied most frequently is educational research. It is from this perspective that the following six areas of application and concomitant examples should be viewed.

DESIGN OF EXPERIMENTS

Although Osburn (1967) has demonstrated algebraically that multiple matrix sampling is the preferred design for a variety of experiments, the procedure has additional merits in the area of experimental design beyond those which can be described mathematically. The following two examples illustrate this point and show how easily multiple matrix sampling may be substituted for the matched-items design in an experiment.

Piper (1972) assessed the end-of-program performance of kindergarten students over a 34-item universe using multiple matrix sampling. His design involved two groups of students—103 students who had participated in an experimental music program and 96 students who had participated in the music program provided by the school district. Each item in the universe was designed to be administered individually and mastery of each concept was assessed at three levels of difficulty. Piper selected multiple matrix sampling because (a) comprehensive assessment was essential, (b) administering all 34 items would have been very tiring to individual kindergarten students (item performance would have been more indicative of fatigue than ability), and (c) it was necessary to keep the total testing time within limits acceptable to the school district. The 34-item universe was divided through stratified-random sampling (stratification being on the difficulty level of the item) into 6 subtests, 4 containing 6 items and 2 containing 5 items. Each subtest was administered to approximately 15 students selected randomly from the experimental program and to the same number selected from the district program. Piper elected to administer one subtest to every student in both the experimental group and the control group because the total number of students involved was small. (The students participating in the district program selected for testing were those who had been matched originally on the basis of pretest scores with the experimental group.) Noteworthy in this example are two realities, namely, that the age of the students frequently limits the number of test items which can be administered to each student and that most testing programs must operate within some predetermined or acceptable time limits. Multiple matrix sampling is suited ideally to both realities and in this example was the only feasible alternative available. Additionally, Piper noted that the examiners found the test administration procedure less boring because different tests (i.e., subtests) were being administered to different students.

The second example is focused on the assessment of language proficiency in a bilingual Spanish/English language program designed for students in the elementary grades. In the Spring of 1972, the Santa Paula School District, Santa Paula, California field-tested the Michigan Oral Language Series (Standard English as a second language) using approximately 300 elementary grade students. In the initial phase of the project a 600-item universe was developed which measured in depth every objective taught within the program. Half of the items in the universe were in Spanish; the other half, in English. Items within each of these two major content strata were further subdivided into three difficulty levels according to the syntactic structure of both the question and the minimally acceptable answer. The item universe was divided into 20 subtests containing 30 items each using a stratified-random sampling plan. The performance of the 300 students over the 600-item universe was estimated both at the beginning and end of the program through multiple matrix sampling. At each testing, 5 subtests were selected randomly for each of the 10 participating classes and each subtest was administered to 3 students within each class. Over classes, either 6 or 9 students responded to each subtest with the total number of students tested on each occasion numbering approximately 150. In assigning subtests to classes, it should be noted that several plans could have been used. For example, one alternative plan would have been to assign randomly two subtests to each class with from 7 to 8 students responding to each subtest. In this plan, the ability of each class would have been reflected by 60 items and, what is more important, their performance could have been a function of the specific items included in the subtests. In place of this, 5 subtests totalling 150 items were used within each class to minimize this problem and to expose each class to a broad item base. In both cases, however, approximately the same number of students respond to each subtest over classes. When multiple matrix sampling is used for pretesting and posttesting, the same subtests may be used on both occasions. In this example, the same 20 subtests used at the pretest were assigned randomly to classes at posttest and the students tested were also selected randomly. On those few occasions when the same subtest was assigned to the same student both pre and post, the student was automatically given a different posttest.

The Santa Paula assessment of bilingual performance illustrates well one of the prime advantages of multiple matrix sampling—the ability to come to grips with reasonably large item universes. It has always been unreasonable to evaluate an entire instructional program through student performance on any subset of, say, 30, 50, or 100 items and the instances in which this is done should decline sharply in the future as multiple matrix sampling becomes accepted more widely.

QUESTIONNAIRES AND SURVEYS

One of the major themes of this book is that there is absolutely no reason for thinking that multiple matrix sampling is applicable solely to achievement tests. The procedure is applicable generally to measurement instruments and the research of Burton and Remer (1972) and Loadman (1972) demonstrates how the procedure is extended to questionnaires.

Burton and Remer were concerned primarily with determining the feasibility of using multiple matrix sampling in assessing responses to a 74-item questionnaire three pages in length. Five forms of the questionnaire were constructed with each form sent to 100 students selected randomly from the student body. Forms I, II, and III of the questionnaire each consisted of approximately 25 items and were, respectively, the first three pages of the questionnaire; Form IV contained 25 items selected through stratified-random sampling from the first three pages, and; Form V, contained the entire 74-item questionnaire. Contrary to what might have been expected, the return rate for each form did not differ significantly. This is of interest because one reason for using multiple matrix sampling with questionnaires distributed through the mail is the belief that a shorter form will be returned more frequently than a longer one. It may be that college students as a group are well trained in completing forms and the five forms of the questionnaire may have been merely "another form" to them. However, it should be noted that this result is in line with those of two other investigations (e.g., Sletto, 1950, and Champion and Sear, 1969) concerning questionnaire length and response rate. Burton and Remer also examined the responses to those questions included in more than one form and found that students responded similarly to a particular question regardless of the subquestionnaire in which it was found—suggesting that any "context effect" may be minimal.

While the investigation by Burton and Remer was concerned primarily with exploring multiple matrix sampling, Loadman (1972) used the procedure out of necessity. The task faced by him was determining the opinions held by two groups of teachers—student teachers and classroom teachers—about an innovative elementary school program. Because the number of required questions equalled 193 and the estimated completion time for this many questions was two hours, Loadman built two forms of the questionnaire. Thirteen questions (from the 193) were included in both forms and the remaining 180 questions were divided randomly between the two forms. Each form of the questionnaire contained 103 questions. The two forms of the questionnaire were alternated among teachers as they were distributed to the assembled groups. An analysis of the responses to the 13 items common to both forms revealed that, for both groups of teachers, the responses to the questions when found in the first form did not differ significantly from those obtained from the second form. Loadman interpreted this result to mean that any context effect due to form was minimal and that the responses to both forms could be combined reasonably to estimate the responses of all teachers (either student teachers or classroom teachers) to the 193 questions. Loadman, stating originally that a 193-item questionnaire would be very tiring to the teachers, noted that " . . . a fatigue factor began to enter the picture as the subjects completed the one hour of information gathering, thus reinforcing this researcher's personal satisfaction with the sample procedure."

Using multiple matrix sampling to estimate the parameters of questionnaires makes it possible for the researcher to (a) reduce the time commitment required by each participant in completing the questionnaire, and (b) increase the number and variety of questions in the total questionnaire. Although it may be the case that, when the questionnaire is to be mailed, a shorter form is not returned with a significantly greater rate than a longer form, a shorter form is frequently more acceptable when the questionnaire is being administered to an assembled group. It should be noted that questions within questionnaires are interrelated frequently (e.g., If "No" on Question 13, go to Question 20) and complications such as these must be incorporated in constructing subquestionnaires.

AFFECTIVE MEASUREMENT

Multiple matrix sampling has been used with two instruments designed to measure affect. The first considered here was reported by Shoemaker (1971) and involved measuring the degree of affect to six stimuli when items were in paired-comparisons format; the second, reported by Pugh (1971), deals with 4-point Likert items (4=strongly agree, 3=agree, 2=disagree, and 1=strongly disagree).

Shoemaker investigated the feasibility of using multiple matrix sampling with an instrument whose items were in paired-comparisons format. In this particular test, each item consisted of a pair of pictures depicting classroom activities and the student was asked to indicate which of the two activities he preferred to do. The design used by Shoemaker was post mortem item-examinee sampling with the data base consisting of responses made by 407 primary grade students to a 15-item test. In postmortem sampling, the researcher has the responses of all students to all items but acts as if only certain students had responded to certain items. One advantage of postmortem sampling is that the values of all parameters are known in advance.

In the method of paired-comparisons, the preliminary analysis of the data involves the construction of a *P*-matrix which, for *s* stimuli, is an *s* by *s* matrix with cell entries denoting the proportion of examinees who judge the column stimulus more favorable than the row stimulus. When multiple matrix sampling is used here, the proportion of times stimulus *i* is preferred over stimulus *j* among those examinees comparing *i* and *j* is used to estimate the *ij* cell in the *P*-matrix. Several item-examinee sampling plans were used and the results of each plan were replicated 25 times. The conclusion drawn by Shoemaker was that scale values determined by the method of paired-comparisons can be approximated satisfactorily through multiple matrix sampling.

Pugh (1971) was interested primarily in determining the relative merits of item-examinee sampling and examinee-sampling using Likert items. In his postmortem investigation, the data base consisted of the responses of 600 sixth-grade students to 60 Likert items. Various item-examinee samples and examinee samples were taken by Pugh and the results from each sample were used to estimate the mean score and the variance of scores (the score for each student was the sum of his ratings over items) for the 600 students. The conclusion drawn by Pugh was that such parameters could be estimated satisfactorily through item-examinee sampling and, concomitantly, that item-examinee sampling is generally more accurate than examinee-sampling.

STATEWIDE ASSESSMENT

Although multiple matrix sampling will have an impact on many areas of research in the future, in educational research its most immediate and notable impact will be in the area of statewide assessment. There is an increasing awareness on the part of many states that any improvement in the educational system must be based on an accurate and detailed assessment of the performance levels of its students in a variety of content areas. The piecemeal testing programs of earlier years are being replaced appropriately by an educational assessment program built upon a comprehensive set of agreed-upon goals for each content area. Two states which have adopted multiple matrix sampling for the assessment of student performance at the state level are California and Florida. A general description of the manner in which multiple matrix sampling is being used in Florida for statewide assessment is given by Christian (1971) and a detailed explanation of how California used the procedure to assess mathematics achievement is given by Tardif (1971) and Riles (1971). The latter two serve well as an example of multiple matrix sampling used for statewide assessment and are considered here in more detail.

In 1971, the Office of Evaluation within the California Department of Education assessed mathematics achievement in grades three, six, and eight. Initially, item universes were developed for each grade by committees charged specifically with this task. The resulting item universe for third-grade mathematics consisted of 528 items; for grade 6, 952 items, and; for grade 8, 1,105 items. Nineteen objectives were covered by the universe for grade 3, 27 for grade 6, and 28 for grade 8. All items were pretested before being used.

The item universes were subdivided subsequently into subtests each containing from 9 to 12 items resulting in 54, 98 and 115 subtests, respectively, for grades 3, 6, and 8. Three hundred schools were selected randomly from the state for each of the three grade levels and one class from each of the schools which agreed to participate was selected randomly for testing. (It should be noted here that this procedure for sampling examinees is known more commonly as "cluster sampling" and is not the same as sampling examinees randomly from the total population of examinees. The relative merits of cluster sampling are discussed by Lord [1959]). Not all schools selected initially agreed to participate and, of those who did participate, not all sent their results to Tardif before the deadline. The result was that data were received from 211 third grades, 223 sixth grades, and 219 eighth grades. Concerning any potential bias here, Tardif (1971) states that " . . . the director of this . . . testing has no reason to believe that the sample of classes was atypical of the population of all classes at the grade level of interest." The subtests were group-administered and all students within each class were tested, with each student receiving a different subtest. The assembled test package was mailed to each class and after testing all materials were mailed back to the Office of Program Evaluation. In the final analysis, 5,596 grade 3 students, 6,419 grade 6 students, and 6,125 grade 8 students were tested.

In discussing the benefits of using multiple matrix sampling in conjunction with an item universe (as opposed to the traditionally-administered tests containing a small number of items), Tardif makes the point most strongly that the latter approach provides only a limited basis for bringing about program changes while the former provides a substantial amount of very specific information. The results of the 1971 testing, which pinpoint the strengths and deficiencies in the mathematics learned by students in California, "will be utilized by the Department of Education in shaping recommendations for future thrusts in mathematics education to include guideline suggestions for future textbook adoptions."

One last point should be made here and is this. The item universes used for each grade level were not developed inexpensively. This is merely a reflection of the fact that constructing an item universe is time-consuming and involves several steps, i.e., selecting the objectives, empaneling the content committee, and pretesting and revising items. (It cost the State of California approximately $100,000 to produce the 1,500 mathematics items used in its statewide testing. The cost per item was approximately $67.) Although the development cost is initially high, the same basic item universe may be used on successive years requiring only minor modifications to certain items or groups of items as experience with the universe dictates. When averaged over years of use, the initial development cost of an item universe is more tolerable.

EVALUATING THE EFFECTIVENESS OF COMPETING INSTRUCTIONAL PROGRAMS

One problem with which educational researchers are faced continually is evaluating the effectiveness of competing instructional programs. The problem is not resolved easily because the majority of instructional programs available currently have been devised by content experts and each merits some consideration. Although at an early time the goal may have been to find the one "good" program from among the other "bad" programs, a more reasonable goal today is to determine the relative advantages and disadvantages of each program under consideration. Only when the strengths and limitations of each program have been determined accurately are educators in a reasonable position to select one program from among several which best fulfills their requirements.

Multiple matrix sampling used in conjunction with a relevant item universe provides an excellent procedure for examining the relative merits of competing instructional programs. This approach to the problem stems from recognition that the primary measurement requirement in evaluating competing programs is group assessment and not the assessment of individual differences among students. The key feature in this approach is developing an item universe which spans broadly the content domain under consideration. Such an item universe will almost always include more content and more objectives than that considered within any individual program. Here, a "fair" assessment of a program is taken as that assessment procedure which provides the maximum opportunity for students within each program to demonstrate everything they have learned through contact with that program. To accomplish this, the evaluation procedure should attempt to assess the broadest possible spectrum of objectives and content. While few would disagree with this sentiment, the critic may argue that it is a plea for testing everything, implying that such is impossible. This point is an extreme extension of what is meant here. It is possible to construct an item universe for a particular content domain whose items cover those objectives considered by most discipline specialists to be important. In this approach, it is student performance on the item universe which becomes the standard against which all programs are measured and that which provides the basis for interprogram comparisons. A more detailed discussion of this point as well as an examination of alternative procedures for comparing instructional programs is given by Shoemaker (1972).

ESTIMATION OF COVARIANCE AND CORRELATION MATRICES

Item and test covariance matrices (and, hence, correlation matrices) may be estimated through multiple matrix sampling. However, a sampling plan for items or tests is required such that all possible pairs of items or tests are included in one or more subtests or sub-batteries. For example, consider estimating the entries in a covariance matrix for a 5-item test. To compute the covariance of item 1 with item 2, there must be a subgroup of examinees responding to both items. The same procedure must be followed for items 1 and 3, 1 and 4, 1 and 5, 2 and 3, 2 and 4, 2 and 5, 3 and 4, 3 and 5, and 4 and 5; however, the same examinee subgroup need not respond to all possible item pairs. A test covariance matrix is determined similarly with items being replaced by tests. A procedure such as this sets the stage for multiple matrix sampling playing an important role in a variety of multivariate procedures as, for example, factor analysis. Although this application as yet has not received much attention, some important preliminary developments have been reported by Olkin and Pratt (1958), Lord (1960), Ray, Hundleby, and Goldstein (1962), Knapp (1968), Timm (1970), and Peterson and Anderson (1971). In pursuing this development, the analyses reported by Carroll (1961), Ruben (1966) and Binder (1959) should be of use.

Chapter Seven

Prospects for the Future

The applications of multiple matrix sampling have just begun to be explored and the reader can expect many new and exciting developments in the future. Although such applications are not confined to education, it is perhaps in this area that multiple matrix sampling will be taken advantage of most readily. An increasing number of administrators, researchers, teachers, and citizens are recognizing the necessity of a comprehensive assessment of all educational systems. What is required, of course, is a carefully constructed item universe for each instructional program of interest. It is because such universes are beginning to assume an important role in education that multiple matrix sampling will be used—and be required—more frequently.

The applications of multiple matrix sampling are both numerous and ubiquitous. If a potential data base may be cast as an object-variable matrix, multiple matrix sampling may be used to estimate parameters of interest. Indeed, one future application of the procedure will be estimating parameters of data bases having more than two dimensions. For example, one three-dimensional matrix of interest in education might be a student (row) by items (columns) by instructional program (layers) matrix representing the achievement of a group of students during an entire year of study. When considered in the future, the two-dimensional model will undoubtedly be the case in which multiple matrix sampling is applied most simply.

Many research problems lie ahead in multiple matrix sampling. Some have been mentioned in this book and others will become apparent as the procedure is extended to new and different areas of application. The researcher interested in pursuing these topics can anticipate many trials and tribulations. It should be noted, however, that what is worthwhile is seldom achieved easily.

Appendixes

Appendix A

Program for Estimating Parameters through Multiple Matrix Sampling

1.0 - PROGRAM IDENTIFICATION

Program for Estimating Parameters through Multiple Matrix Sampling

2.0 - OBJECTIVE

This program estimates selected parameters and standard error of estimate associated with each parameter given multiple matrix sampling.

3.0 - PROGRAM DESCRIPTION

Using item scores sequenced by subtest and, within each subtest, by examinee as input, the program computes an estimate of each parameter for each subtest and then pools results over subtests to provide the best estimate of each parameter. The parameters estimated within the program are: (a) mean universe score, (b) second central moment of universe scores, (c) third central moment, (d) fourth central moment, (e) component of variance for items, (f) component of variance for examinees, (g) component of variance for item x examinee interaction, (h) average item reliability, (i) item difficulty index for each item, and (j) reliability of the universe scores. The equations used to estimate parameters (a) through (d) were derived from Hooke (1954, 1956); the equations for estimating components of variance are given in numerous sources (e.g., Myers, 1966); and the equation for estimating the average item reliability is given by Winer (1971). A pooled (over subtests) estimate of each parameter as well as the estimated standard error associated with each parameter is accomplished by the jackknife procedure. A description of the jackknife procedure is given by Shoemaker (1972). The 90, 95, 97.5, and 99 percent confidence intervals for paramcters (a) through (h) are computed using the *t*-distribution (df = number of subtests minus one) and jackknifed standard errors of estimate. Additionally, the standard error for the mean and variance of universe scores is computed using the Hooke-Lord-Novick procedure given in detail by Pandey and Shoemaker (1973). The reliability of the universe scores is computed by projecting the pooled average item reliability to the total universe score by the Spearman-Brown Prophesy formula. Theoretical boundaries for the second through fourth central moments of universe scores are determined through a linear programming procedure described by Hill (1972) and any estimate not within bounds is flagged before setting it equal to the midpoint of its acceptable limits. Using estimates of moments, the normative relative frequency distribution is approximated by the negative hypergeometric distribution, the lognormal distribution and/or one of the family of Pearson distributions as selected by the user. The equation for the negative hypergeometric distribution is given by Lord and Novick (1968); equations for the lognormal distributions, by Aitchison and Brown (1957); and, equations for the family of Pearson distributions, by Elderton and Johnson (1969).

4.0 - SUBROUTINES AND FUNCTIONS

4.1 - Coded Subprograms

BMOMTS - Determination of minimum and maximum values for each estimated central moment.

SUBROUTINE BMOMTS (V1,U2,U3,U4,KPOP)

V1 - Estimated mean universe score
U2 - Estimated variance of universe scores
U3 - Estimated third central moment
U4 - Estimated fourth central moment
KPOP - Number of items in universe

FRV - Evaluation of $\cos(x)^r e^{-\nu x} dx$ from -1.570796 to 1.570796 using Simpson's rule.

FUNCTION FRV (R,VV)

R - 6.*(B2-B1-1.)/(2.*B2-3.*B1-6.)
VV - SQRT((R*R*(R-2.)**2*B1)/(16.*(R-1.)-B1*(R-2.)**2))

GAMMA - Computation of gamma function using the recursion relation and polynomial approximation.

FUNCTION GAMMA (XX)

XX - Any real argument

JKNIFE - Calculation of pooled estimates of parameters and associated standard error of estimate per parameter by the jackknife procedure.

SUBROUTINE JKNIFE (U1,U2,U3,U4)

U1 - Pooled estimate of mean universe score
U2 - Pooled estimate of second central moment
U3 - Pooled estimate of third central moment
U4 - Pooled estimate of fourth central moment

LOAD - Duplication of data base on scratch file number 2. No arguments.

LOGNL - Graduation of normative distribution with lognormal distribution.

SUBROUTINE LOGNL (KPOP,MITEM,XBAR,XVAR)

KPOP - Number of items in universe
MITEM - Maximum item score
XBAR - Mean of test score distribution
XVAR - Variance of test scores

LP - Linear programming subroutine. (Subroutine written originally by Dr. James E. Martin, Virginia Polytechnic Institute, and modified subsequently by Dr. Richard K. Hill, Jr., Virginia Polytechnic Institute, and David M. Shoemaker.)

SUBROUTINE LP (NERR)

NERR - Error code return

MOMVAL - Conversion of frequences in *X*-matrix from alphanumeric to numeric and calculation of moment value. (Subroutine written originally by Dr. Richard K. Hill, Jr., Virginia Polytechnic Institute, and modified subsequently by David M. Shoemaker.)

SUBROUTINE MOMVAL (VAL,MOM,VALC)

VAL - Value of moment about origin
MOM - Degree of moment
VALC - Value of moment about mean

NEGHGR - Graduation of normative distribution by negative hypergeometric distribution.

SUBROUTINE NEGHGR (KPOP,XBAR,XVAR)

KPOP - Number of items in universe
XBAR - Mean universe score
XVAR - Variance of universe scores

PEARSN - Graduation of normative distribution with one of the family of Pearson curves.

SUBROUTINE PEARSN (U1,U2,U3,U4,L1,L2,STEP,NPCUR

U1 - Mean universe score
U2 - Second central moment
U3 - Third central moment
U4 - Fourth central moment
L1 - Lower limit of index
L2 - Upper limit of index
STEP - Desired step size in going from L1 to L2
NPCURV - Specific code number of curve to be selected by program if user wants to override curve selected normally by subroutine.

POLRT – Computation of the real and complex roots of a real polynomial.

SUBROUTINE POLRT (XCOF,M,ROOTR,ROOTI,IER)

XCOF - Vector of $M + 1$ coefficients of the polynomial ordered from smallest to largest power

M - Order of polynomial

ROOTR - Resultant vector of length M containing real roots

ROOTI - Resultant vector of length M containing the corresponding imaginary roots

IER - Error code return

RDFMT - Subroutine for inputting variable format.

SUBROUTINE RDFMT (FMT,N)

FMT - Vector containing format

N - Number of format cards inputted

SETUP – Preparation of X-matrix for use by LP subroutine. (Subroutine written originally by Dr. Richard K. Hill, Jr., Virginia Polytechnic Institute)

SUBROUTINE SETUP (NI,NJ,NK,XMULT)

NI - Degree of moment

NJ - Maximum possible universe score plus seven

NK - Degree of moment

XMULT - Switch value indicating which boundary (minimum or maximum) is to be computed

SUBTST - Calculation of parameter estimates for each subtest.

SUBROUTINE SUBTST (KPOP,NDISK)

KPOP - Number of items in universe

NDISK - Data base storage

0 = no storage on scratch file

1 = data base stored on scratch file

TTABLE – Calculation of Student's t corresponding to prescribed alpha level (two-tailed probability).

SUBROUTINE TTABLE (T,N,AREA)

T - Student's *t*
N - Degrees of freedom associated with *t*
AREA - Relative area associated with confidence interval

VCOV - Calculation of variance of item-test covariances for subtest

SUBROUTINE VCOV (XX,NSPT,IPT,VCOVIT)

XX - Matrix containing item scores for each examinee
NSPT - Number of examinees (rows of XX)
IPT - Number of items (columns of XX)
VCOVIT - Variance of item-test covariances

4.2 - Library Subprograms

ALOG10	EXIT
ALOG	DSQRT
ABS	
SQRT	
ATAN	
COS	
EXP	

5.0 - DATA SPECIFICATIONS

5.1 - Input Formats

Input formats are described in detail at the beginning of the program listing. An example of an input data structure is given in Attachment 1.

5.2 - Output Formats

Output resulting from the input data structure listed in Attachment 1 is given in Attachment 2.

6.0 - PROGRAM CONSTRAINTS AND LIMITATIONS

6.1 - Programming Language
Fortran IV

6.2 - Storage Requirements
Total core for execution is 300K
See 7.0 for scratch file requirements

6.3 - Hardware Configuration
IBM 360 Model 91, Punched cards

6.4 - Program Parameters
6.4.1 - Item scoring: items may be scored either dichotomously or polychotomously.
6.4.2 - The number of subtests within any data structure is limited to 100.
6.4.3 - The number of items per subtest is limited to 150.

6.5 - Error Messages
The following error messages may be generated by the program:

1. Errors on card specifying problem set (Note: error terminates program)	(MAIN)
2. Error in *t*-table arg: df = _	(TTABLE)
3. Error in *t*-table arg: p = _	(TTABLE)
4. Excessive number of format cards (Note: error terminates program)	(RDFMT)
5. Invalid limits and/or delta arguments	(PEARSN)
6. Ordinate undefined at this point	(PEARSN)
7. M .LE.1. for Type XI	(PEARSN)
8. M out of range for Type VIII	(PEARSN)
9. M negative for Type IX	(PEARSN)
10. Gamma arg within .0000001 of being a negative integer	(GAMMA)
11. Gamma arg GT 57, overflow, ans set to 1.E75	(GAMMA)
12. No solution for_____moment	(LP)
13. Distribution aborted A21 LE 0 or GT 1	(NEGHGR)
14. Incorrect code for Pearson curve specified by user. Curve will be selected by program	(PEARSN)

7.0 - OPERATING INSTRUCTIONS

Submit the program in batch mode in the following form:

col 1
// job card
// PASSWORD

```
// EXEC FORTGCLG,REGION=300K
//   FORT.SYSIN DD *

  Insert Main and Subroutines here.

/*
// GO.FT02F001 DD DSN=&SCRATCH,UNIT=SYSDA,
// DCB=(RECFM=FBA,BLKSIZE=7280,LRECL=80),
// SPACE=(TRK,150,RLSE)
// GO.SYSIN DD *

  Insert data structures here.

/*
//
```

8.0 PROGRAM FLOW CHART (MACRO)

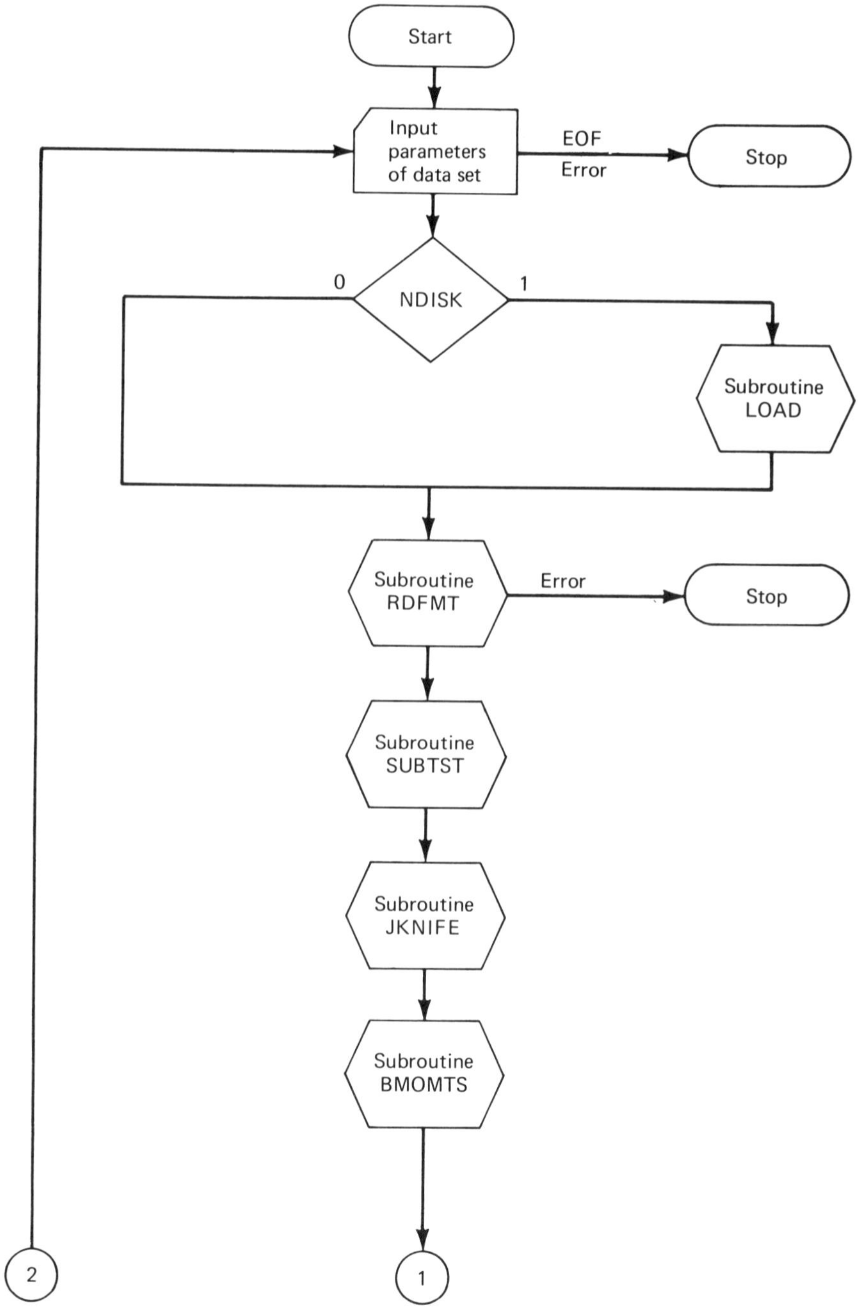

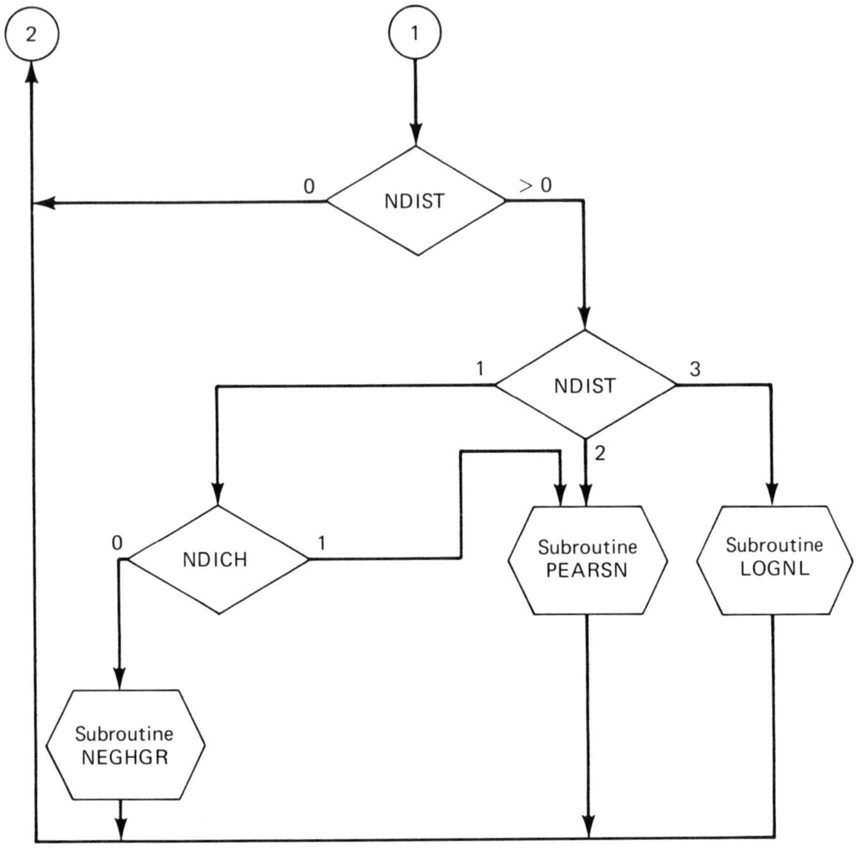
2
1
0
NDIST
> 0
1
NDIST
3
2
0
NDICH
1
Subroutine PEARSN
Subroutine LOGNL
Subroutine NEGHGR

```
FORTRAN IV G LEVEL 21            MAIN            DATE = 73?01         07/06/07            PAGE 0001
C
C  DATE OF MOST RECENT REVISION: 19 FEB 73
C***********************************************************************
C
C  ESTIMATION OF PARAMETERS OF UNIVERSE SCORES THROUGH MULTIPLE MATRIX
C  SAMPLING
C
C  DAVID M. SHOEMAKER
C  SOUTHWEST REGIONAL LABORATORY FOR EDUCATIONAL
C     RESEARCH AND DEVELOPMENT
C
C
C  ORGANIZATION OF DATA STRUCTURES
C
C     1.  DATA CARDS ARE ORGANIZED SEQUENTIALLY INTO PROBLEM SETS.  EACH
C         PROBLEM SET CONTAINS (IN THIS ORDER) THE FOLLOWING CARD SETS:
C
C          TITLE CARD
C          PARAMETER CARD
C          DATA BLOCK FOR SUBTEST 1
C          DATA BLOCK FOR SUBTEST 2
C          ...
C          DATA BLOCK FOR SUBTEST T
C
C
C     2.  ORGANIZATION OF INDIVIDUAL CARD SETS:
C
C
C            CARD SET                   COLUMN    DESCRIPTION
C                                             (INTEGERS RIGHT-JUSTIFIED)
C            -------------------        ------    ----------------------
C
C        A.  TITLE CARD (1 CARD)        01-72     ALPHANUMERIC TITLE OF
C                                                 PROBLEM SET
C
C        B.  PARAMETER CARD (1 CARD)    01-03     NUMBER OF ITEMS
C                                                 IN UNIVERSE
C
C                                       04-05     MAX ITEM SCORE
C
C                                       06-08     NUMBER OF SUBTESTS
C
C                                         10      METHOD OF ITEM SCORING
C
C                                                      0=DICHOTOMOUS
C                                                      1=NON-DICHOTOMOUS
C
C                                         12      NORMATIVE DISTRIBUTION
C
C                                                      0=NONE DESIRED
C                                                      1 = ESTIMATE DIST.
C                                                          (CURVE PICKED
C                                                          BY PROGRAM)
C
```

```
C                                         2 = PEARSON CURVE
C
C                                         3 = LOGNORMAL
C                                             CURVE
C
C                              16-30     MEAN
C                              31-45     VARIANCE
C                              46-60     3RD CENTRAL MOMENT
C                              61-75     4TH CENTRAL MOMENT
C
C                              NOTE: THIS PROGRAM MAY ALSO BE
C                              USED TO GRADUATE NORMATIVE
C                              DISTRIBUTIONS WITH THE NECESSARY
C                              MOMENTS CALCULATED IN ADVANCE.
C                              IN THIS CASE, THE REQUIRED
C                              MOMENTS ARE INPUTTED ON COLUMNS
C                              16 TO 75 (WITH DECIMAL POINT
C                              PUNCHED ON THE CARD).  THE NUMBER
C                              OF SUBTESTS (COLUMNS 6 TO 8)
C                              SHOULD BE SET EQUAL TO 0.  IF
C                              ESTIMATES OF PARAMETERS ARE TO
C                              BE COMPUTED THROUGH MULTIPLE
C                              MATRIX SAMPLING, IGNORE COLUMNS
C                              16 THROUGH 75.
C
C                              79-80     CODE NUMBER OF SPECIFIC
C                                        PEARSON CURVE TO BE
C                                        USED.  IF LEFT BLANK,
C                                        CURVE WILL BE SELECTED
C                                        BY PROGRAM. THE
C                                        FOLLOWING CODES MAY BE
C                                        USED:
C
C                                        01 = TYPE I CURVE
C                                        02 = TYPE II CURVE
C                                        03 = TYPE III CURVE
C                                        ...
C                                        11 = TYPE XI CURVE
C                                        12 = NORMAL
C
C
C
C     C.  DATA BLOCK FOR SUBTEST I: THE DATA BLOCK FOR EACH SUBTEST
C         CONSISTS OF THREE CARD SETS.
C
C          1)  PARAMETER CARD (1 CARD)
C          2)  FORMAT CARD SET (K CARDS)
C          3)  DATA CARDS CONTAINING ITEM SCORES PER EXAMINEE
C              SEQUENCED BY EXAMINEE
C
C
C          THE ORGANIZATION OF THESE CARD SETS IS AS FOLLOWS:
C
C
```

```
FORTRAN IV G LEVEL  21              MAIN              DATE = 73101        07/06/07              PAGF 0003

C         PARAMETER CARD      01-04  NUMBER OF EXAMINEES IN SUBGROUP
C
C                             05-08  NUMBER OF ITEMS PER SUBGROUP
C
C                             09-12  NO. OF CARDS FOR EACH EXAMINEE
C
C         FORMAT CARD SET     01-72  STANDARD FORTRAN IV FORMAT
C                                    PUNCHED IN COLUMNS 01-72 ON
C                                    EACH CARD AND ENCLOSED IN
C                                    PARENTHESES FOR INPUTTING ITEM
C                                    SCORES FOR EACH EXAMINEE IN THE
C                                    ITEM-EXAMINEE SAMPLE.  THE
C                                    NUMBER OF CARDS CONTAINING THE
C                                    FORMAT MAY NOT EXCEED NINE FOR
C                                    EACH ITEM-EXAMINEE DATA SET.
C                                    THE FIRST CARD AFTER THE FORMAT
C                                    CARDS MUST CONTAIN "END OF
C                                    FORMAT" IN COLUMNS 01-13.
C
C         DATA CARD SET              THE ITEM SCORES OF EACH
C                                    EXAMINEE ON SUBTEST SEQUENCED
C                                    BY EXAMINEE AND ORGANIZED ON
C                                    CARDS ACCORDING TO FORMAT
C                                    SPECIFICATION
C
C     3.  THE PROGRAM WILL PROCESS REPEATED PROBLEM SETS.
C
C     4.  RESTRICTIONS:
C
C        A.  NUMBER OF SUBTESTS LIMITED TO 100
C        B.  NUMBER OF ITEMS PER SUBTEST LIMITED TO 150
C        C.  MAXIMUM TEST SCORE (NO. OF ITEMS TIMES MAXIMUM ITEM SCORE)
C            ON UNIVERSE MAY NOT EXCEED 999.
C             (RESTRICTION MODIFIED EASILY, HOWEVER.)
C
C
C     5.  IT IS FREQUENTLY THE CASE THAT THE INVESTIGATOR WISHES TO
C         ASSEMBLE ITEM SCORES INTO CONTENT STRATA AND TO PROCESS THE
C         ITEMS WITHIN EACH CONTENT STRATUM AS THOUGH THEY WERE ITEM
C         POPULATIONS PER SE.  CONSIDER, FOR EXAMPLE, THE CASE IN WHICH
C         HALF THE ITEMS IN EACH SUBTEST MEASURE OUTCOME CASE A WITH THE
C         OTHER HALF MEASURING OUTCOME CASE B.  HERE, THE INVESTIGATOR
C         MAY WANT TO ANALYZE RESULTS SEPARATELY FOR A AND B.  THIS
C         COULD BE HANDLED THROUGH DATA STRUCTURES 1. AND 2. BUT TWO
C         DUPLICATE SETS OF ITEM SCORES WOULD BE REQUIRED.  TO AVOID
C         USING DUPLICATE SETS OF ITEM SCORES, IT IS POSSIBLE WITHIN
C         THE PROGRAM TO STORE A DATA SET ON A SCRATCH FILE AND THEN
C         MAKE REPEATED PASSES THROUGH IT USING DIFFERENT FORMAT CARDS
C         (OF THE TYPE DESCRIBED IN 2C).
C
C
C         THE ORGANIZATION OF INPUT CARDS FOR THIS CAPABILITY IS ONLY
C         SLIGHTLY DIFFERENT FROM THE NORMAL CARD ORGANIZATION. ASSUME
C         FOR EXAMPLE, THAT THERE ARE 5 PROBLEM SETS EACH OF WHICH IS TO
```

```
      C          USE THE SAME DATA BASE. WITHIN THE DATA STREAM, THIS DATA BASE
      C          PRECEEDS THE PROBLEM SETS.  THE DATA BASE MUST BE PRECEEDED BY
      C          A CARD WITH "STR2" PUNCHED IN COLS. 1 THROUGH 4 AND TERMINATED
      C          BY A CARD WHICH HAS "END" PUNCHED IN COLUMNS 1 THROUGH 3.
      C          EACH PROBLEM SET WHICH IS TO USE THIS DATA BASE MUST HAVE A
      C          "1" PUNCHED IN COLUMN 14 OF CARD 2B. WHEN THIS OPTION IS USED,
      C          THE "DATA CARD SET" DESCRIBED IN 2C IS OMITTED.  EVERYTHING
      C          ELSE STAYS THE SAME.
      C
      C              A.  PROBLEM SETS USING THIS STORAGE FEATURE MAY BE
      C                  INTERSPERSED AMONG STANDARD PROBLEM SETS.
      C
      C              B.  A CALL TO STORE A DATA BASE MAY BE INSERTED AT ANY
      C                  PLACE (PRECEEDING A PROBLEM SET) WITHIN THE THE DATA
      C                  STREAM.  A DATA BASE MAY BE STORED ANY NUMBER OF
      C                  TIMES, BUT ONLY ONE BASE IS IN STORAGE AT ANY
      C                  PARTICULAR TIME.
      C
      C              C.  SCRATCH FILE 2 IS USED WITHIN THE PROGRAM.  THE
      C                  STORAGE CAPACITY OF THIS FILE IS 13,650 CARD IMAGES.
      C                  IF MORE STORAGE SPACE IS NEEDED, AN ADDITIONAL FILE
      C                  MUST BE USED.
      C
      C              D.  IF THE DATA ARE STORED ON THE SCRATCH FILE (I.E., THE
      C                  'STR2' OPTION IS USED) AND IF, IN MAKING A PASS
      C                  THROUGH THE DATA, THE RESULTS OF A PARTICULAR SUBTEST
      C                  ARE TO BE IGNORED COMPLETELY BECAUSE IT CONTAINS NO
      C                  ITEMS FOR A PARTICULAR STRATUM OF INTEREST, ONLY THE
      C                  PARAMETER CARD DESCRIBED IN SECTION 2C IS USED.  IN
      C                  THIS CASE, COLUMNS 01-04 CONTAIN THE NUMBER OF CARD
      C                  IMAGES TO BE IGNORED AND COLUMNS 05-08 ARE LEFT BLANK.
      C
      C*********************************************************************
      C
0001        COMMON /BLOCK1/ SDATA(102,11),NTEST
0002        DATA STR2/4HSTR2/
0003        DIMENSION TITLE(18)
      C
      C  READ SPECIFICATIONS FOR PROBLEM SET
      C
0004   1000 READ (5,1,END=5000) (TITLE(I),I=1,18)
0005        IF ( TITLE(1) .NE. STR2 ) GO TO 50
0006        CALL LOAD
0007        GO TO 1000
0008   50   READ (5,7) KPOP,MITEM,NTEST,NDICH,NDIST,NDISK,U1,U2,U3,U4,NPCURV
0009        IF ( NDISK .EQ. 1 ) REWIND 2
      C
0010        WRITE (6,2) (TITLE(I),I=1,18)
0011        IF ( U1 .LE. 0. ) WRITE (6,3) KPOP,MITEM,NTEST,NDICH,NDIST
0012        IF ( U1 .GT. 0. ) WRITE (6,4) KPOP,MITEM,NDIST,U1,U2,U3,U4
0013        IF ( NDISK .GT. 0 ) WRITE (6,9)
      C
      C  CHECK ON PARAMETERS
      C
```

```
FORTRAN IV G LEVEL  21              MAIN              DATE = 73101          07/06/07              PAGE 0005

0014              NERR=0
0015              IF ( NDICH .LT. 0 .OR. NDICH .GT. 1 ) NERR=NERR+1
0016              IF ( NDIST .LT. 0 .OR. NDIST .GT. 4 ) NERR=NERR+1
0017              IF ( KPOP .LE. 0 ) NERR=NERR+1
0018              IF ( U1 .NE. 0. .AND. NTEST .GT. 0 ) NERR=NERR+1
0019              IF ( NERR .LE. 0 ) GO TO 90
0020              WRITE (6,5) NERR
0021              CALL EXIT
0022         90   CONTINUE
            C
            C  BYPASS ESTIMATION PROCEDURE IF MOMENTS ARE AVAILABLE
            C
0023              IF ( U1 .GT. 0. ) GO TO 100
            C
            C  ESTIMATE PARAMETERS FROM SUBTEST RESULTS
            C
0024              CALL SUBTST (KPOP,NDISK)
            C
            C  COMPUTE POOLED ESTIMATE (OVER SUBTESTS) OF PARAMETER AND STANDARD
            C  ERROR OF ESTIMATE ASSOCIATED WITH PARAMETER THROUGH JACKKNIFE
            C  PROCEDURE
            C
0025              CALL JKNIFE(U1,U2,U3,U4)
            C
            C  ESTIMATION OF RELIABILITY OF UNIVERSE SCORE
            C
0026              THETA=SDATA(101,8)
0027              RTT=KPOP*THETA/(1.+KPOP*THETA)
0028              WRITE (6,8) RTT
            C
            C  TEST FEASIBILITY OF ESTIMATED MOMENTS USING LINEAR PROGRAMMING:
            C  MOMENTS FOUND TO BE OUT OF BOUNDS ARE CORRECTED WITHIN BMOMTS
            C
0029        100   MAXT=KPOP*MITEM
0030              IF ( MAXT .LE. 1000 ) CALL BMOMTS (U1,U2,U3,U4,MAXT)
0031              IF ( MAXT .GT. 1000 ) WRITE (6,10)
            C
            C  GRADUATE NORMATIVE DISTRIBUTION
            C
0032              IF ( NDIST .EQ. 0 ) GO TO 1000
0033              S=MAXT
0034              GO TO (110,120,130), NDIST
0035        110   IF ( NDICH .EQ. 0 ) CALL NEGHGR (KPOP,U1,U2)
0036              IF ( NDICH .EQ. 1 ) CALL PEARSN (U1,U2,U3,U4,0.,S,1.,NPCURV)
0037              GO TO 1000
0038        120   CALL PEARSN (U1,U2,U3,U4,0.,S,1.,NPCURV)
0039              GO TO 1000
0040        130   CALL LOGNL (KPOP,MITEM,U1,U2)
0041              GO TO 1000
            C
            C  EXIT GRACEFULLY
            C
0042       5000   WRITE (6,6)
0043              CALL EXIT
```

```
FORTRAN IV G LEVEL  21                MAIN            DATE = 73101          07/06/07              PAGE 0006

              C
              C  FORMAT STATEMENTS
              C
 0044         1     FORMAT (18A4)
 0045         2     FORMAT (1H1,18A4//)
 0046         3     FORMAT (31H SPECIFICATIONS FOR PROBLEM SET//
                  119H NUMBER OF ITEMS    I11/
                  219H MAX ITEM SCORE     I11/
                  319H NUMBER OF SUBTESTSI11/
                  413H ITEM SCORINGI17/
                  523H NORMATIVE DISTRIBUTIONI7)
 0047         4     FORMAT (31H SPECIFICATIONS FOR PROBLEM SET//
                  119H NUMBER OF ITEMS    I11/
                  219H MAX ITEM SCORE     I11/
                  323H NORMATIVE DISTRIBUTIONI7/
                  44H U1'E16.7/3H U2E17.7/3H U3E17.7/3H U4E17.7)
 0048         5     FORMAT (4H ***I3,38H ERRORS ON CARD SPECIFYING PROBLEM SET//)
 0049         6     FORMAT (20H ALL INPUT PROCESSED)
 0050         7     FORMAT (I3,I2,I3,3I2,1X,4F15.0,I5)
 0051         8     FORMAT (//45H ESTIMATED RELIABILITY OF UNIVERSE SCORE IS  F8.5//)
 0052         9     FORMAT (17H STR2 OPTION USED)
 0053         10    FORMAT (//51H TEST FOR FEASIBILITY OF ESTIMATED MOMENTS SKIPPED:/
                  137H MAX UNIVERSE SCORE EXCEEDS 1000      //)
 0054               END
```

```
FORTRAN IV G LEVEL  21                MAIN            DATE = 73101         07/06/07            PAGE 0007

 *OPTIONS IN EFFECT*  ID,EBCDIC,SOURCE,NOLIST,NODECK,LOAD,NOMAP
 *OPTIONS IN EFFECT*  NAME = MAIN    , LINECNT =        56
 *STATISTICS*    SOURCE STATEMENTS =        54,PROGRAM SIZE =      2584
 *STATISTICS*  NO DIAGNOSTICS GENERATED
```

```
FORTRAN IV G LEVEL  21                BLK DATA          DATE = 73101          07/06/07               PAGE 0001

0001                 BLOCK DATA
0002                 INTEGER A,B,C,D,E,F,G,H
0003                 COMMON /BLOCK8/ A(135),B(135),C(135),D(135),E(21),F(55),
                    1     G(55),H(135),I(135),J(135),K(135),L(21)
0004                 DATA A/
                    1 1, -6, -6,  3,  3,  8,  8, -6, -6, 24,  6,  6,-12,-12,-24,
                    2-24,  6, -6, -6,-24,-24, 36, 36, 24, 24, 48,  3, 16,-18,-18,
                    3-48,-48, 36,  1,  0, -1,  0, -2,  0,  2,  0, -4, -1, -1,  4,
                    4  2,  6,  4, -2,  2,  1,  6,  4,-12, -6, -6, -8,-12, -1, -4,
                    5  6,  6, 16, 12,-12,  1,  0, -1,  0, -2,  0,  2, -4, -1, -1,
                    6  2,  4,  4,  6, -2,  1,  2,  4,  6, -6,-12, -8, -6,-12, -1,
                    7 -4,  6,  6, 12, 16,-12,  1,  0,  0,  0, -1,  0,  0,  0,  0,
                    8 -4,  0,  0,  0,  2, -2,  0,  0,  0,  6,  0,  0,  8,  0,  1,
                    9  0, -3, -6, -8,  0,  6,  1,  0,  0,  0, -1,  0,  0,  0,  0/
0005                 DATA B/
                    1 -4,  0,  0,  2,  0, -2,  0,  0,  0,  6,  8,  0,  0,  1,  0,
                    2 -6, -3,  0, -8,  6,  1,  0, -1,  0,  0,  0,  0,  0,  0, -3,
                    3  0,  0,  0,  0, -3,  0,  6,  0,  3,  0,  6,  0,  2, -3,  0,
                    4 -8, -6,  6,  1,  0, -1,  0,  0,  0,  0,  0,  0, -3,  0,  0,
                    5  0,  0, -3,  0,  6,  0,  3,  6,  0,  2,  0, -3, -6, -8,  6,
                    6  1,  0,  0,  0,  0,  0,  0,  0,  0,  0,  0,  0,  0,  0, -6,
                    7  0,  0,  0,  0,  0,  0,  3,  0,  8,  0, -6,  1,  0,  0,  0,
                    8  0,  0,  0,  0,  0,  0,  0,  0,  0,  0, -6,  0,  0,  0,  0,
                    9  0,  0,  3,  0,  8, -6,  1,  0,  0, -1, -1, -1, -1,  1,  0/
0006                 DATA C/
                    1  0, -1, -1,  2,  2,  2,  2,  3,  0,  1, -2, -2, -4, -4,  4,
                    2  1,  0,  0,  0, -2, -2,  0, -1, -1,  0,  0,  2,  2,  2,  2,
                    3  4,  1,  0, -2, -2, -4, -4,  4,  1,  0,  0,  0,  0,  0, -1,
                    4 -1, -2, -2,  2,  2,  2,  2,  2,  1,  2, -2, -2, -4, -4,  4,
                    5  1,  0,  0,  0, -1,  0,  0,  0,  0, -1,  0,  0, -2,  0,  0,
                    6  0,  1,  2,  2,  0, -2,  1,  0,  0, -1,  0,  0,  0,  0,  0,
                    7 -1, -2,  0,  0,  0,  0,  2,  1,  0,  2, -2,  1,  0,  0,  0,
                    8  0,  0,  0, -1,  0, -1,  0, -2,  0,  0,  1,  0,  2,  2, -2,
                    9  1,  0,  0,  0,  0,  0,  0, -1,  0, -1, -2,  0,  0,  0,  1/
0007                 DATA D/
                    1  2,  2, -2,  1,  0,  0,  0,  0,  0,  0,  0,  0,  0,  0,  0,
                    2 -1, -1,  0,  0,  1,  1,  0,  0,  0, -1,  0,  0, -2,  0, -1,
                    3  0,  1,  2,  2,  0, -2,  1,  0,  0,  0, -1, -2,  0,  0, -1,
                    4  0,  2,  1,  0,  2, -2,  1,  0, -1,  0, -1,  0, -1,  0, -1,
                    5  1,  0,  2,  2, -2,  1,  0, -1,  0, -1, -1,  0, -1,  0,  1,
                    6  2,  2, -2,  1,  0,  0,  0,  0,  0,  0, -1,  0, -2,  0,  2,
                    7  1,  0,  0,  0,  0,  0,  0, -1,  0, -2,  2,  1,  0,  0,  0,
                    8  0, -1,  0,  0, -1,  1,  1,  0,  0,  0,  0, -1, -1,  0,  1,
                    9  1,  0,  0,  0,  0, -1, -1,  1,  1,  0, -1, -1,  0,  0,  1/
0008                 DATA E/
                    1  1,  0,  0, -1, -1,  1,  1,  0,  0,  0, -1,  1,  0,  0, -1,
                    2  1,  0, -1,  1, -1,  1/
0009                 DATA F/
                    1  1, -3, -3,  2,  2,  6,  3, -6, -6,  4,  1,  0, -1,  0, -2,
                    2 -1,  3,  2, -2,  1,  0, -1, -2, -1,  2,  3, -2,  1,  0,  0,
                    3  0, -3,  0,  2,  1,  0,  0,  0, -3,  2,  1,  0, -1, -1,  1,
                    4  1, -1, -1,  1,  1,  0, -1,  1, -1,  1/
0010                 DATA G/
                    1  1, -1,  1, -1,  0,  1,  2, -3,  0,  1,
```

```
FORTRAN IV G LEVEL  21              BLK DATA           DATE = 73101         07/06/07              PAGE 0002
                    2  2,  0, -3,  0,  1,  1, -1, -1,  0,  0,
                    3  1,  1, -1, -1,  0,  0,  0,  1, -2,  3,
                    4  2, -1,  0, -2, -1,  1, -2,  2,  3,  0,
                    5 -1, -2, -1,  0,  1,  4, -6, -6,  2,  2,
                    6  6,  3, -3, -3,  1/
0011                 DATA H/
                    1  1, -1,  1, -1,  0,  1,  1, -2,  0,  1,  1,  0, -2,  0,  1,
                    2  2, -3,  0,  0,  0,  1,  2,  0, -3,  0,  0,  0,  1, -6, 12,
                    3  0, -3,  0, -4,  0,  1, -6,  0, 12,  0, -3,  0, -4,  0,  1,
                    4  1, -1, -1,  0,  0,  0,  0,  0,  0,  1,  1, -1, -1,  0,  0,
                    5  0,  0,  0,  0,  0,  1,  1, -1, -1,  0,  0,  0,  0,  0,  0,
                    6  0,  0,  1, -1,  2,  1, -1,  0,  0,  0,  0,  0, -2,  0,  0,
                    7  1, -1,  1,  2,  0, -1,  0,  0,  0,  0, -2,  0,  0,  0,  1,
                    8 -2,  3,  2,  0,  0, -1,  0,  0,  0, -2, -1,  0,  0,  0,  1,
                    9 -2,  2,  3,  0,  0,  0, -1,  0,  0, -2, -1,  0,  0,  0,  0/
0012                 DATA I/
                    1  1,  1, -2, -2,  1,  1,  0,  0,  0,  0,  4,  0,  0, -2, -2,
                    2  0,  0,  1, -1,  2,  1, -1,  0,  0,  0,  0,  0,  0, -1, -1,
                    3  0,  0,  0,  0,  0,  1, -1,  1,  2,  0, -1,  0,  0,  0,  0,
                    4  0, -1, -1,  0,  0,  0,  0,  0,  0,  1, -2,  3,  2,  0,  0,
                    5 -1,  0,  0,  0, -2,  0, -1,  0,  0,  0,  0,  0,  0,  0,  1,
                    6 -2,  2,  3,  0,  0,  0, -1,  0,  0, -2,  0, -1,  0,  0,  0,
                    7  0,  0,  0,  0,  0,  1,  6,-12, -6,  3,  0,  4,  0, -1,  0,
                    8  8,  2,  2, -2,  0, -2,  0,  0, -1,  0, -2,  0,  1,  6, -6,
                    9-12,  0,  3,  0,  4,  0, -1,  8,  2,  2,  0, -2,  0, -2,  0/
0013                 DATA J/
                    1  0, -1,  0, -2,  0,  1,  2, -3, -4,  0,  2,  1,  0,  0,  0,
                    2  4,  1,  1,  0, -2, -1,  0,  0,  0, -1, -1,  0,  0,  0,  1,
                    3  2, -4, -3,  2,  0,  0,  1,  0,  0,  4,  1,  1, -2,  0,  0,
                    4 -1,  0, -1,  0,  0, -1,  0,  0,  0,  1,  4, -6, -6,  0,  0,
                    5  2,  2,  0,  0,  6,  2,  1,  0,  0, -2, -2,  0,  0,  0, -1,
                    6 -1,  0,  0,  0,  0,  1,  1, -2, -2,  1,  1,  0,  0,  0,  0,
                    7  0,  2,  2,  0,  0,  0,  0,  0, -2, -2,  0,  0,  0,  0,  0,
                    8  0,  0,  1,  4, -6, -6,  0,  0,  2,  2,  0,  0,  6,  0,  3,
                    9  0,  0,  0,  0,  0,  0,  0, -3, -3,  0,  0,  0,  0,  0,  0/
0014                 DATA K/
                    1  1, -6, 12, 12, -3, -6, -4,  0,  1,  0,-16, -4, -4,  4,  8,
                    2  4,  0, -2,  2,  4,  4,  0, -2,  0, -4,  0,  0, -1,  0,  1,
                    3 -6, 12, 12, -6, -3,  0, -4,  0,  1,-16, -4, -4,  8,  4,  0,
                    4  4, -2,  4,  2,  0,  4,  0, -2,  0, -4,  0, -1,  0,  0,  1,
                    5-12, 24, 18, -6,  0, -8, -6,  2,  0,-24, -6, -6,  6,  0,  6,
                    6  6,  0,  3,  0,  6,  6, -3,  0,  0, -3, -3,  0, -1,  0,  0,
                    7  1,-12, 18, 24,  0, -6, -6, -8,  0,  2,-24, -6, -6,  0,  6,
                    8  6,  6,  0,  0,  3,  6,  6,  0, -3, -3,  0, -3,  0, -1,  0,
                    9  0,  0,  1, 36,-72,-72, 18, 18, 24, 24, -6, -6, 96, 24, 24/
0015                 DATA L/
                    1-24,-24,-24,-24,  6,-12,-12,-24,-24, 12, 12, 12, 12, 12,  3,
                    2  4, -3, -3, -4, -4,  1/
0016                 END
```

```
FORTRAN IV G LEVEL  21              BLK DATA        DATE = 73101        07/06/07          PAGE 0003

  *OPTIONS IN EFFECT*  ID,EBCDIC,SOURCE,NOLIST,NODECK,LOAD,NOMAP
  *OPTIONS IN EFFECT*  NAME = BLK DATA, LINECNT =       56
  *STATISTICS*  NO DIAGNOSTICS GENERATED
```

```
FORTRAN IV G LEVEL  21              BMOMTS              DATE = 73101          07/06/07              PAGE 0001

0001              SUBROUTINE BMOMTS (V1,U2,U3,U4,KPOP)
            C
            C  DETERMINATION OF MINIMUM AND MAXIMUM VALUES FOR EACH ESTIMATED
            C  CENTRAL MOMENT
            C
0002              COMMON /BLOCK5/ SSUM(4),X(10,1007),B(6,4),W(6)
0003              DIMENSION C(4,7)
            C
            C  CONVERT CENTRAL MOMENTS TO MOMENTS ABOUT ORIGIN
            C
0004              V2=U2+V1*V1
0005              V3=U3+3.*V1*V2-2.*V1**3
0006              V4=U4+4.*V3*V1-6.*V2*V1*V1+3.*V1**4
0007              SSUM(1)=V1
0008              SSUM(2)=V2
0009              SSUM(3)=V3
0010              SSUM(4)=V4
0011              DO 40 I=1,4
0012              DO 40 J=1,7
0013         40   C(I,J)=0.
0014              DO 50 I=1,4
0015         50   C(I,1)=SSUM(I)
0016              C(1,4)=0.
0017              C(2,4)=U2
0018              C(3,4)=U3
0019              C(4,4)=U4
0020              X(1,5)=7
0021              X(1,6)=KPOP+1
0022              NJ=KPOP+7
            C
            C  COMPUTE BOUNDARIES FOR MOMENTS ABOUT ORIGIN AND CONVERT TO
            C  BOUNDARIES FOR CENTRAL MOMENTS
            C
0023              WRITE (6,5)
0024              WRITE (6,1)
0025         90   DO 100 I=4,6
0026              MOM=I-2
0027              CALL SETUP (I,NJ,I,1.)
0028              CALL LP(NERR)
0029              IF ( NERR .EQ. 0 ) GO TO 110
0030              CALL MOMVAL (C(MOM,2),MOM,C(MOM,5))
0031              CALL SETUP (I,NJ,I,-1.)
0032              CALL LP (NERR)
0033              CALL MOMVAL (C(MOM,3),MOM,C(MOM,6))
0034              C(MOM,7)=0.
0035              IF ( C(MOM,1) .LT. C(MOM,2) .OR. C(MOM,1) .GT. C(MOM,3) )GO TO 110
0036        100   CONTINUE
0037              GO TO 115
0038        110   C(MOM,7)=1.
            C
            C
0039        115   IF ( C(3,5) .LE. C(3,6) ) GO TO 130
0040              TEMP=C(3,5)
0041              C(3,5)=C(3,6)
```

```
      C(3,6)=TEMP
  130 CONTINUE
C
C  PRINTOUT BOUNDARIES FOR MOMENTS AND APPRAISE ESTIMATED MOMENTS
C
      I=1
      WRITE (6,4) I,C(1,1),C(1,4)
      DO 150 I=2,4
      IF ( C(I,1) .LT. C(I,2) .OR. C(I,1) .GT. C(I,3) ) GO TO 145
      WRITE (6,2) I,(C(I,J),J=1,6)
      GO TO 150
  145 WRITE (6,3) I,(C(I,J),J=1,6)
  150 CONTINUE
C
C  FIXUP BAD MOMENT ESTIMATES BY CHANGING ESTIMATED BAD MOMENT
C  TO MEAN OF ESTIMATED BOUNDARIES
C
      WRITE (6,5)
      IF ( C(2,7) .EQ. 0. ) GO TO 160
      V2=(C(2,2)+C(2,3))/2.
      SSUM(2)=V2
      C(2,1)=V2
      C(2,4)=V2-V1*V1
      WRITE (6,6)
      GO TO 90
  160 IF ( C(3,7) .EQ. 0. ) GO TO 170
      V3=(C(3,2)+C(3,3))/2.
      SSUM(3)=V3
      C(3,1)=V3
      C(3,4)=V3-3.*V1*V2+2.*V1**3
      WRITE (6,6)
      GO TO 90
  170 IF ( C(4,7) .EQ. 0. ) GO TO 200
      C(4,1)=(C(4,2)+C(4,3))/2.
      C(4,4)=(C(4,5)+C(4,6))/2.
      WRITE (6,6)
      I=4
      WRITE (6,4) I,C(4,1),C(4,4)
  200 U2=C(2,4)
      U3=C(3,4)
      U4=C(4,4)
      RETURN
    1 FORMAT (60H MINIMUM AND MAXIMUM POSSIBLE VALUES FOR ESTIMATED MOME
     1NTS  //23X,12HABOUT ORIGIN,35X,10HABOUT MEAN/
     210X,38H--------------------------------------,
     3 7X,38H--------------------------------------//
     41X,6HMOMENT,3X,38HESTIMATE        MINIMUM        MAXIMUM,
     57X,         38HESTIMATE        MINIMUM        MAXIMUM
     69X,9HAPPRAISAL//)
    2 FORMAT (1X,I3,3X,6E15.7,8X,2HOK)
    3 FORMAT (1X,I3,3X,6E15.7,8X,6HNOT OK)
    4 FORMAT (1X,I3,3X,E15.7,30X,E15.7)
    5 FORMAT (///)
    6 FORMAT (   31H ATTEMPTED FIXUP OF BAD MOMENTS///)
```

```
FORTRAN IV G LEVEL  21                BMOMTS          DATE = 73101        07/06/07          PAGE 0003
 0083                END
```

```
FORTRAN IV G LEVEL  21              BMOMTS          DATE = 73101        07/06/07          PAGE 0004

 *OPTIONS IN EFFECT*  ID,EBCDIC,SOURCE,NOLIST,NODECK,LOAD,NOMAP
 *OPTIONS IN EFFECT*  NAME = BMOMTS  , LINECNT =       56
 *STATISTICS*    SOURCE STATEMENTS =       83,PROGRAM SIZE =     2700
 *STATISTICS*  NO DIAGNOSTICS GENERATED
```

```
FORTRAN IV G LEVEL  21                FRV              DATE = 73101         07/06/07              PAGE 0001

0001                 FUNCTION FRV (R,VV)
               C
               C  EVALUATION OF COS(X)**R*EXP(-V*X)DX FROM -PI/2 TO PI/2 USING
               C  SIMPSON'S RULE
               C
0002                 V=ABS(VV)
0003                 PI=3.141593
0004                 D=PI/100.
0005                 FRV=0.
0006                 DO 100 I=1,99,2
0007                 T=-1.570796+I*D
0008            100  FRV=FRV+2.*COS(T)**R*EXP(-V*T)+COS(T+D)**R*EXP(-VT-VD)
0009                 FRV=2.*D*FRV/3.
0010                 RETURN
0011                 END
```

```
FORTRAN IV G LEVEL  21              FRV              DATE = 73101          07/06/07            PAGE 0002

  *OPTIONS IN EFFECT*  ID,EBCDIC,SOURCE,NOLIST,NODECK,LOAD,NOMAP
  *OPTIONS IN EFFECT*  NAME = FRV     , LINECNT =       56
  *STATISTICS*    SOURCE STATEMENTS =        11,PROGRAM SIZE =      736
  *STATISTICS*  NO DIAGNOSTICS GENERATED
```

```
FORTRAN IV G LEVEL  21                GAMMA           DATE = 73101        07/06/07           PAGE 0001

0001                FUNCTION GAMMA (XX)
           C
           C  COMPUTATION OF GAMMA FUNCTION USING THE RECURSION RELATION AND
           C  POLYNOMIAL APPROXIMATION
           C
0002                IF ( XX-57. ) 6,6,4
0003          4     WRITE (6,2)
0004                GAMMA=1.E75
0005                RETURN
0006          6     X=XX
0007                ERR=1.E-6
0008                GX=1.
0009                IF ( X-2. ) 50,50,15
0010         10     IF ( X-2. ) 110,110,15
0011         15     X=X-1.
0012                GX=GX*X
0013                GO TO 10
0014         50     IF ( X-1. ) 60,120,110
           C
           C  TEST FOR X NEAR NEGATIVE INTEGER OR ZERO
           C
0015         60     IF ( X-ERR ) 62,62,80
0016         62     Y=FLOAT(INT(X))-X
0017                IF ( ABS(Y)-ERR ) 130,130,64
0018         64     IF ( 1.-Y-ERR ) 130,130,70
           C
           C  X NOT NEAR A NEGATIVE INTEGER OR ZERO
           C
0019         70     IF ( X-1. ) 80,80,110
0020         80     GX=GX/X
0021                X=X+1.
0022                GO TO 70
0023        110     Y=X-1.
0024                GY=1.+Y*(-.5771017+Y*(.9858540+Y*(-.8764218+Y*(.8328212+Y*
                   1(-.5684729+Y*(.2548205+Y*(-.05149930)))))))
0025                GAMMA=GX*GY
0026        120     RETURN
0027        130     WRITE (6,1)
0028                RETURN
0029          1     FORMAT (/53H GAMMA ARG WITHIN .000001 OF BEING A NEGATIVE INTEGER/
                   1)
0030          2     FORMAT (/44H GAMMA ARG GT 57, OVERFLOW, ANS SET TO 1.E75/)
0031                END
```

```
FORTRAN IV G LEVEL  21                GAMMA            DATE = 73101          07/06/07            PAGE 0002

  *OPTIONS IN EFFECT*  ID,EBCDIC,SOURCE,NOLIST,NODECK,LOAD,NOMAP
  *OPTIONS IN EFFECT*  NAME = GAMMA    , LINECNT =       56
  *STATISTICS*    SOURCE STATEMENTS =        31,PROGRAM SIZE =     1036
  *STATISTICS*  NO DIAGNOSTICS GENERATED
```

```
FORTRAN IV G LEVEL  21                 JKNIFE           DATE = 73101          07/06/07              PAGE 0001

0001                 SUBROUTINE JKNIFE (U1,U2,U3,U4)
               C
               C  CALCULATION OF POOLED ESTIMATE OF PARAMETER AND ASSCCIATED STANDARD
               C  ERROR OF ESTIMATE BY THE JACKKNIFE PROCEDURE
               C
               C  MAP FOR SDATA MATRIX
               C
               C     SDATA (J,1) TO SDATA(J,8)    PARAMETER ESTIMATES FROM SUBTEST J
               C     SDATA(J,9)                   NO. OF EXAMINEES FOR SUBTEST J
               C     SDATA(J,10)                  NO. OF ITEMS FOR SUBTEST J
               C     SDATA(J,11)                  NO. OF OBSERVATIONS FOR SUBTEST J
               C     SDATA(101,I)                 POOLED ESTIMATE FOR PARAMETER I
               C     SDATA(102,I)                 ESTIMATED SE FOR PARAMETER I
               C
0002                 COMMON /BLOCK1/ SDATA(102,11),NTEST
0003                 DIMENSION A(8),B(8),PRCNT(6),CONF(6)
0004                 DATA CONF/4H70.0,4H80.0,4H90.0,4H95.0,4H97.5,4H99.0/
0005                 DATA PRCNT/.70,.80,.90,.95,.975,.99/
               C
0006                 IF ( NTEST .LE. 1 ) RETURN
               C
0007                 DO 200 I=1,8
               C
               C  CALCULATION OF WEIGHTED PARAMETER ESTIMATE USING ALL SUBTEST RESULTS
               C
0008                 SDATA(101,I)=0.
0009                 SDATA(102,I)=0.
0010                 NGOOD=0
0011                 PALL=0.
0012                 TOBS=0.
0013                 DO 100 J=1,NTEST
0014                 IF ( SDATA(J,I) .EQ. -9999999. ) GO TO 100
0015                 SDATA(J,11)=SDATA(J,9)*SDATA(J,10)
0016                 TOBS=TOBS+SDATA(J,11)
0017                 PALL=PALL+SDATA(J,I)*SDATA(J,11)
0018                 NGOOD=NGOOD+1
0019            100  CONTINUE
0020                 IF ( NGOOD .EQ. 0 ) GO TO 200
0021                 IF ( NGOOD .GT. 1 ) GO TO 125
0022                 SDATA(101,I)=PALL/TOBS
0023                 GO TO 200
               C
               C  CALCULATION OF PSEUDOVALUES PER SUBTEST FOR A GIVEN PARAMETER
               C
0024            125  SP=0.
0025                 SPP=0.
0026                 DO 150 J=1,NTEST
0027                 IF ( SDATA(J,I) .EQ. -9999999. ) GO TO 150
0028                 P=NGOOD*PALL/TOBS-(NGOOD-1.)*(PALL-SDATA(J,I)*SDATA(J,11))/
                    1     (TOBS-SDATA(J,11))
0029                 SP=SP+P
0030                 SPP=SPP+P*P
0031            150  CONTINUE
               C
```

```
FORTRAN IV G LEVEL  21              JKNIFE           DATE = 73101         07/06/07                PAGE 0002

              C  CALCULATION OF POOLED ESTIMATE OF PARAMETER AND ESTIMATE OF STANDARD
              C  ERROR OF ESTIMATE ASSOCIATED WITH PARAMETER
              C
 0032               SDATA(101,I)=SP/NGOOD
 0033               T= (SPP-SP*SP/NGOOD)/(NGOOD*(NGOOD-1.))
 0034               IF  ( T .GT. 0. )  SDATA(102,I)=SQRT(T)
              C
              C
 0035           200 CONTINUE
              C
              C
 0036               WRITE (6,1)  ((SDATA(J,I),I=1,8),J=101,102)
              C
              C
              C  COMPUTE CONFIDENCE INTERVALS FOR PARAMETERS
              C  (DF=NUMBER OF SUBTESTS MINUS ONE)
              C
 0037               DO 300 J=3,6
 0038               CALL TTABLE (T,NTEST-1,PRCNT(J))
 0039               WRITE (6,2) CONF(J)
 0040               DO 250 I=1,8
 0041               A(I)=0.
 0042               B(I)=0.
 0043               IF  ( SDATA(102,I) .EQ. 0. ) GO TO 250
 0044               TEMP=T*SDATA(102,I)
 0045               A(I)=SDATA(101,I)-TEMP
 0046               B(I)=SDATA(101,I)+TEMP
 0047           250 CONTINUE
 0048               WRITE (6,3)  (A(K),K=1,8),(B(K),K=1,8)
 0049           300 CONTINUE
              C
 0050               U1=SDATA(101,1)
 0051               U2=SDATA(101,2)
 0052               U3=SDATA(101,3)
 0053               U4=SDATA(101,4)
              C
 0054               RETURN
              C
 0055            1  FORMAT (//5H MEAN8E15.5/5H SE  8E15.5//)
 0056            2  FORMAT (1X,A4,1X,27HPERCENT CONFIDENCE INTERVAL)
 0057            3  FORMAT (5H FROM8E15.5/5H TO  8E15.5//)
 0058               END
```

```
FORTRAN IV G LEVEL  21                JKNIFE          DATE = 73101        07/06/07              PAGE 0003

  *OPTIONS IN EFFECT*  ID,EBCDIC,SOURCE,NOLIST,NODECK,LOAD,NOMAP
  *OPTIONS IN EFFECT*  NAME = JKNIFE  , LINECNT =       56
  *STATISTICS*    SOURCE STATEMENTS =       58,PROGRAM SIZE =      1944
  *STATISTICS*  NO DIAGNOSTICS GENERATED
```

```
FORTRAN IV G LEVEL  21                LOAD            DATE = 73101        07/06/07            PAGE 0001

 0001                    SUBROUTINE LOAD
                 C
                 C  DUPLICATION OF DATA BASE ON SCRATCH FILE NUMBER 2
                 C
 0002                    DIMENSION X(20)
 0003                    DATA END/3HEND/
 0004                    REWIND 2
 0005                    WRITE (6,2)
 0006             100    READ (5,1)  (X(I),I=1,20)
 0007                    WRITE (6,3)  (X(I),I=1,20)
 0008                    IF ( X(1) .EQ. END ) GO TO 200
 0009                    WRITE (2,1)  (X(I),I=1,20)
 0010                    GO TO 100
 0011             200    END FILE 2
 0012                    RETURN
 0013              1     FORMAT (20A4)
 0014              2     FORMAT (//37H1CARD IMAGES LOADED ON SCRATCH FILE 2//)
 0015              3     FORMAT (1X,20A4)
 0016                    END
```

```
FORTRAN IV G LEVEL  21                LOAD              DATE = 73101          07/06/07              PAGE 0002

  *OPTIONS IN EFFECT*  ID,EBCDIC,SOURCE,NOLIST,NODECK,LOAD,NOMAP
  *OPTIONS IN EFFECT*  NAME = LOAD    , LINECNT =       56
  *STATISTICS*    SOURCE STATEMENTS =       16,PROGRAM SIZE =      704
  *STATISTICS*  NO DIAGNOSTICS GENERATED
```

```
FORTRAN IV G LEVEL  21              LOGNL              DATE = 73101          07/06/07              PAGE 0001

0001                SUBROUTINE LOGNL (KPOP,MITEM,XBAR,XVAR)
              C
              C  GRADUATION OF NORMATIVE DISTRIBUTION WITH LOGNORMAL DISTRIBUTION
              C
0002                DIMENSION Y(1000)
0003                PI=3.141593
0004                S=KPOP*MITEM
              C
              C  ESTIMATION OF DIRECTION OF SKEWNESS
              C
              C     1 = POSITIVELY-SKEWED
              C     2 = NEGATIVELY-SKEWED
              C
0005                NDIST=1
0006                IF ( XBAR .GT. S/2. ) NDIST=2
              C
              C  COMPUTATION OF CONSTANTS FOR LOGNORMAL DISTRIBUTION
              C
0007                IF ( NDIST .EQ. 2 ) XBAR=S-XBAR
0008                YVAR=ALOG(XVAR/(XBAR*XBAR)+1.)
0009                YBAR=ALOG(XBAR)-YVAR/2.
0010                CLN1=1./SQRT(2.*PI*YVAR)
0011                CLN2=2.*YVAR
              C
0012                WRITE (6,2) YBAR,YVAR
              C
              C  CALCULATION OF ORDINATES
              C
0013                WRITE (6,3)
0014                K=0
0015                T=0.
0016           100  IF ( T .GT. S ) GO TO 150
0017                IF ( T .GT. 0. ) GO TO 125
0018                ORD=0.
0019                GO TO 130
0020           125  ORD=(CLN1/T)*EXP(-((ALOG(T)-YBAR)**2)/CLN2)
0021           130  K=K+1
0022                Y(K)=ORD
0023                ABSC=T
0024                IF ( NDIST .EQ. 2 ) ABSC=S-T
0025                WRITE (6,4) ABSC,ORD
0026                T=T+1.
0027                GO TO 100
              C
              C  APPROXIMATION OF AREA UNDER CURVE BY TRAPEZOID RULE
              C
0028           150  AREA=0.
0029                NSTOP=K-1
0030                DO 200 I=2,NSTOP
0031           200  AREA=AREA+Y(I)
0032                AREA=(Y(1)+Y(K))/2.+AREA
              C
0033                WRITE (6,5) AREA
              C
```

```
FORTRAN IV G LEVEL  21              LOGNL            DATE = 73101        07/06/07            PAGE 0002

0034                 RETURN
                C
0035              1  FORMAT (65H1GRADUATION OF NORMATIVE DISTRIBUTION WITH LOGNORMAL DI
                    1STRIBUTION//)
0036              2  FORMAT (//48H CALCULATED PARAMETERS OF LCGNOBMAL DISTRIBUTION//
                    15H MEANE20.7/9H VARIANCEE16.7)
0037              3  FORMAT (//5X,5HINDEX,5X,8HORDINATE//)
0038              4  FORMAT (F10.0,F13.5)
0039              5  FORMAT (//17H APPROXIMATE AREAE15.7)
0040                 END
```

```
FORTRAN IV G LEVEL  21              LOGNL            DATE = 73101         07/06/07           PAGE 0003

  *OPTIONS IN EFFECT*  ID,EBCDIC,SOURCE,NOLIST,NODECK,LOAD,NOMAP
  *OPTIONS IN EFFECT*  NAME = LOGNL    , LINECNT =       56
  *STATISTICS*    SOURCE STATEMENTS =       40,PROGRAM SIZE =     5440
  *STATISTICS*  NO DIAGNOSTICS GENERATED
```

```
FORTRAN IV G LEVEL  21              LP              DATE = 73101          07/06/07             PAGE 0001

0001                 SUBROUTINE LP (NERR)
               C
               C  LINEAR PROGRAMMING SUBROUTINE
               C
               C  SUBROUTINE WRITTEN ORIGINALLY BY DR. JAMES E. MARTIN (VIRGINIA
               C  POLYTECHNIC INSTITUTE) AND MODIFIED BY DR. RICHARD K. HILL, JR
               C  (VIRGINIA POLYTECHNIC INSTITUTE)
               C
0002                 COMMON /BLOCK5/ SSUM(4),X(10,1007),B(6,4),W(6)
0003                 NERR=1
0004                 S=.000001
0005                 NI=X(1,5)
0006                 NJ=X(1,6)+6
0007                 X(2,5)=0.
0008                 DO 30 I=2,NI
0009            30   X(I,6)=X(I,5)
0010                 DO 32 J=7,NJ
0011            32   X(2,J)=X(1,J)
0012                 NM=0
0013                 DO 34 LL=1,4
0014            34   B(2,LL)=X(2,LL)
0015                 DO 50 I=3,NI
0016                 DO 38 J=1,4
0017            38   B(I,J)=X(I,J)
0018            50   CONTINUE
0019                 NIT=0
0020            51   SUMM=0.
0021                 SUMJ=0.
0022                 M=7
0023                 DO 62 J=7,NJ
0024                 DO 52 I=3,NI
0025                 IF ( B(I,2) .NE. X(I,2) )  GO TO 52
0026                 IF ( B(I,3) .NE. X(I,3) )  GO TO 52
0027                 IF ( B(I,4) .NE. X(I,4) )  GO TO 52
0028                 SUMJ=SUMJ+X(I,J)
0029            52   CONTINUE
0030                 IF ( SUMJ .LE. 0. ) GO TO 60
0031                 IF ( J .EQ. 7 ) GO TO 58
0032                 IF ( SUMM-SUMJ ) 54,56,60
0033            54   M=J
0034                 GO TO 58
0035            56   IF ( X(2,M) .LT. X(2,J) )  GO TO 60
0036                 M=J
0037            58   SUMM=SUMJ
0038            60   SUMJ=0.
0039            62   CONTINUE
0040            64   N=3
0041                 NN=NI-1
0042                 DO 68 I=3,NN
0043                 IF ( X(N,M) .LE. S ) GO TO 66
0044                 R1=X(N,6)/X(N,M)
0045                 IF ( X(I+1,M) .LE. S ) GO TO 68
0046                 R2=X(I+1,6)/X(I+1,M)
0047                 IF ( R1 .LE. R2 ) GO TO 68
```

```
FORTRAN IV G LEVEL  21              LP              DATE = 73101          07/06/07            PAGE 0002

 66   N=I+1
 68   CONTINUE
      DO 70 I=1,4
 70   X(N,I)=X(NI+I,M)
      SAVE=X(N,M)
      PIVOT=1./SAVE
      X(N,M)=-1.
      DO 78 I=2,NI
      IF ( X(I,M)+S ) 76,76,72
 72   IF ( X(I,M)-S ) 74,76,76
 74   W(I)=0.
      GO TO 78
 76   W(I)=-X(I,M)*PIVOT
 78   CONTINUE
      X(N,M)=SAVE
      DO 80 J=6,NJ
      TEMP=X(N,J)
      X(N,J)=0.
      DO 80 I=2,NI
      X(I,J)=X(I,J)+W(I)*TEMP
      IF ( J .EQ. M ) X(N,M)=1.
 80   CONTINUE
      DO 82 I=2,NI
      IF ( I .EQ. N ) GO TO 82
      X(I,M)=0.
 82   CONTINUE
      NIT=NIT+1
      NCT=0
      TEST=0.
 86   DO 88 I=3,NI
      IF ( B(I,2) .NE. X(I,2) ) GO TO 88
      IF ( B(I,3) .NE. X(I,3) ) GO TO 88
      IF ( B(I,4) .NE. X(I,4) ) GO TO 88
      TEST=TEST+X(I,6)
      NCT=NCT+1
 88   CONTINUE
      IF ( NIT .EQ. 1 ) GO TO 96
      IF ( NM .NE. M ) GO TO 96
      IF ( TEST1 .GT. TEST ) GO TO 96
 89   NL=NI-2
      WRITE (6,1) NL,NIT
      NERR=0
      GO TO 999
 96   TEST1=TEST
      NM=M
      IF ( NIT .GT. 100 ) GO TO 89
      IF ( NCT .NE. 0 ) GO TO 51
 98   M=7
      DO 106 J=8,NJ
      IF ( X(2,M)+S ) 100,104,104
100   IF ( X(2,J)+S ) 102,106,106
102   IF ( X(2,M) .LT. X(2,J) ) GO TO 106
104   M=J
106   CONTINUE
```

```
      IF ( (X(2,M)+S) .LT. 0. ) GO TO 64
  999 RETURN
    1 FORMAT (//16H NO SOLUTION FOR,I4,7H MOMENT//)
      END
```

```
FORTRAN IV G LEVEL  21              LP              DATE = 73101          07/06/07            PAGE 0004

  *OPTIONS IN EFFECT*  ID,EBCDIC,SOURCE,NOLIST,NODECK,LOAD,NOMAP
  *OPTIONS IN EFFECT*  NAME = LP      , LINECNT =       56
  *STATISTICS*    SOURCE STATEMENTS =       105,PROGRAM SIZE =      2800
  *STATISTICS*  NO DIAGNOSTICS GENERATED
```

```
FORTRAN IV G LEVEL  21            NEGHGR            DATE = 73101         07/06/07              PAGE 0001

0001                 SUBROUTINE NEGHGR (KPOP,XBAR,XVAR)
               C
               C  GRADUATION OF NORMATIVE DISTRIBUTION BY NEGATIVE HYPERGEOMETRIC
               C  DISTRIBUTION
               C
               C  COMPUTATION OF CONSTANTS
               C
0002                 A21=(KPOP/(KPOP-1.))*(1.-XBAR*(KPOP-XBAR)/(KPOP*XVAR))
0003                 WRITE (6,1) XBAR,XVAR,A21
0004                 IF ( A21 .GT. 0. .AND. A21 .LT. 1. ) GO TO 120
0005                 WRITE (6,2)
0006                 RETURN
0007             120 A=(-1.+1./A21)*XBAR
0008                 B=-A-1.+KPOP/A21
0009                 SLOG1=0.
0010                 SLOG2=0.
0011                 C=A+B
0012                 DO 140 I=1,KPOP
0013                 SLOG1=SLOG1+ALOG10(B-I+1.)
0014             140 SLOG2=SLOG2+ALOG10(C-I+1.)
0015                 C=10.**(SLOG1-SLOG2)
               C
               C  GENERATION OF NEGATIVE HYPERGEOMETRIC DISTRIBUTION
               C
0016                 WRITE (6,3)
0017                 S=KPOP
0018                 N3=KPOP+1
0019                 CK=0.
0020                 DO 200 J=1,N3
0021                 K=J-1
0022                 IF ( K .EQ. 0 ) GO TO 160
0023                 SLOG1=0.
0024                 SLOG2=0.
0025                 SLOG3=0.
0026                 SLOG4=0.
0027                 DO 150 I=1,K
0028                 SLOG1=SLOG1+ALOG10(S-I+1.)
0029                 SLOG2=SLOG2+ALOG10(A+I-1.)
0030                 SLOG3=SLOG3+ALOG10(B-I+1.)
0031             150 SLOG4=SLOG4+ALOG10(FLOAT(I))
0032                 HX=C*10.**(SLOG1+SLOG2-SLOG3-SLOG4)
0033                 GO TO 165
0034             160 HX=C
0035             165 CK=CK+HX
0036                 WRITE (6,4) K,HX,CK
0037             200 CONTINUE
               C
0038                 RETURN
               C
0039               1 FORMAT (//82H APPROXIMATION OF NORMATIVE DISTRIBUTION WITH NEGATIV
                    1E HYPERGEOMETRIC DISTRIBUTION//21H ESTIMATED PARAMETERS//5H MEAN,
                    2F25.5/9H VARIANCEF21.5/4H A21F26.5)
0040               2 FORMAT (//38H DISTRIBUTION ABORTED A21 LE 0 OR GE 1//)
0041               3 FORMAT (//5X,5HINDEX,5X,13HRELATIVE FREQ,10X,12HCUM REL FREQ//)
```

```
FORTRAN IV G LEVEL  21              NEGHGR          DATE = 73101        07/06/07          PAGE 0002

 0042          4     FORMAT (1X,I9,F18.5,F22.5)
 0043                END
```

```
*OPTIONS IN EFFECT*  ID,EBCDIC,SOURCE,NOLIST,NODECK,LOAD,NOMAP
*OPTIONS IN EFFECT*  NAME = NEGHGR  , LINECNT =       56
*STATISTICS*    SOURCE STATEMENTS =        43,PROGRAM SIZE =     1856
*STATISTICS*  NO DIAGNOSTICS GENERATED
```

```
FORTRAN IV G LEVEL  21                  PEARSN            DATE = 73101          07/06/07               PAGE 0001

0001              SUBROUTINE PEARSN(U1,U2,U3,U4,L1,L2,STEP,NPCURV)
            C
            C  GRADUATION OF FREQUENCY DISTRIBUTIONS WITH PEARSON CURVES
            C
            C  PRIMARY REFERENCE
            C
            C  ELDERTON, W.P. AND JOHNSON, N.L. SYSTEMS OF FREQUENCY CURVES.
            C  LONDON/CAMBRIDGE UNIVERSITY PRESS/1969.  (PAGES 35-109)
            C
            C  ARGUMENTS OF SUBROUTINE
            C
            C     U1   MEAN
            C     U2   2ND CENTRAL MOMENT (VARIANCE)
            C     U3   3RD CENTRAL MOMENT
            C     U4   4TH CENTRAL MOMENT
            C     L1   LOWER LIMIT OF INDEX
            C     L2   UPPER LIMIT OF INDEX
            C   STEP   DESIRED STEP SIZE IN GOING FROM L1 TO L2
            C
            C  REQUIRED SUBROUTINES
            C
            C   GAMMA  EXP  ABS  SQRT  COS  ATAN
            C
0002              REAL M,M1,M2,L1,L2,LAMBDA
0003              DIMENSION CF(4),RR(4),RI(4)
            C
            C  COMPUTATION OF INDEX OF SKEWNESS (B1) AND DEGREE OF KURTOSIS (B2)
            C  AND CRITERION VALUE USED IN THE SELECTION OF APPROPRIATE PEARSON
            C  CURVE
            C
0004              X0=0.
0005              XN=0.
0006              NHIGH=0
0007              CUMP=0.
0008              PI=3.1415927
0009              TPI=2.*PI
0010              B1=U3*U3/(U2**3)
0011              B2=U4/(U2*U2)
0012              Z1=B1
0013              Z2=B2
0014              CRIT1=(B1*(B2+3.)**2)/(4.*(4.*B2-3.*B1)*(2.*B2-3.*B1-6.))
0015              WRITE (6,14) U1,U2,U3,U4
0016              WRITE (6,1) B1,B2,CRIT1
0017              IF ( U2 .GT. 0. ) GO TO 60
0018              WRITE (6,15)
0019              RETURN
0020         60   IF ( L1 .LT. L2 .AND. STEP .GT. 0. ) GO TO 65
0021              WRITE (6,16)
0022              RETURN
0023         65   CONTINUE
0024              A=4.*B2-3.*B1
0025              B=3.*B1-2.*B2+6.
0026              CRIT2=(A*(10.*B2-12.*B1-18.)**2-B1*(B2+3.)**2*(8.*B2-9.*B1-12.))/
```

```
FORTRAN IV G LEVEL  21              PEARSN          DATE = 73101         07/06/07            PAGE 0002

0027              CRIT3=5.*B2-6.*B1-9.
0028              CRIT4=2.*B2-3.*B1-6.
0029              WRITE (6,23) CRIT2,CRIT3,CRIT4
            C
            C  BYPASS SELECTION PROCEDURE AND USE CURVE SPECIFIED BY USER
            C
0030              IF ( NPCURV .GT. 0 .AND. NPCURV .LT. 13 ) GO TO 75
0031              IF ( NPCURV .EQ. 0 ) GO TO 70
0032              WRITE (6,28) NPCURV
0033              WRITE (6,27)
0034              GO TO 70
0035         75   WRITE (6,28) NPCURV
0036              GO TO (100,450,600,200,500,300,480,700,750,800,850,410), NPCURV
            C
            C
            C  TEST FOR VALUES OF CRITERIA APPROXIMATING THEORETICAL VALUES
            C
0037         70   IF ( CRIT1.LE. .02 .AND. CRIT1.GE. -.02 ) CRIT1=0.
0038              IF ( CRIT1.LE. 1.02 .AND. CRIT1.GE. .98 ) CRIT1=1.
0039              IF ( CRIT2 .GE. -.02 .AND. CRIT2 .LE. .02 ) CRIT2=0.
0040              IF ( CRIT3 .GE. -.02 .AND. CRIT3 .LE. .02 ) CRIT3=0.
0041              IF ( CRIT4 .GT. -.02 .AND. CRIT4 .LT. .02 ) CRIT4=0.
0042              IF ( B1 .GE. 3.98 .AND. B1 .LE. 4.02 ) Z1=4.
0043              IF ( B2 .LE. 3.02 .AND. B2 .GE. 2.98 ) Z2=3.
0044              IF ( B2 .GE. 8.98 .AND. B2 .LE. 9.02 ) Z2=9.
            C
            C  TEST FOR CURVE BEING MAIN TYPE OR TRANSITION TYPE
            C
0045              IF ( CRIT4 .EQ. 0. ) GO TO 600
0046              IF ( Z1 .EQ. 4. .AND. Z2 .EQ. 9. ) GO TO 800
0047              IF ( CRIT2 ) 80,79,80
0048         79   IF ( CRIT1 .LT. 0. .AND. CRIT3 .LT. 0. ) GO TO 700
0049              IF ( CRIT1 .GT. 1. .AND. CRIT4 .GT. 0. ) GO TO 850
0050              IF ( CRIT1.LT.0..AND.CRIT3.GT.0..AND.CRIT4.LT.0. ) GO TO 750
0051         80   IF ( CRIT1) 100,400,90
0052         90   IF ( CRIT1-1.) 200,500,300
            C
            C  CURVE I *FIRST MAIN TYPE*
            C
0053        100   WRITE (6,4)
0054              R=6.*(B2-B1-1.)/(6.+3.*B1-2.*B2)
0055              C=B1*(R+2.)**2+16.*(R+1.)
0056              A12=SQRT(U2*C)/2.
0057              C=R*(R+2.)*SQRT(B1/C)
0058              M1=(R-2.-C)/2.
0059              M2=(R-2.+C)/2.
            C
            C  IF U3 IS POSITIVE, M2 IS POSITIVE ROOT
            C
0060              IF ( U3 .LE. 0. ) GO TO 120
0061              IF ( M2 .GE. M1 ) GO TO 120
0062              TEMP=M2
0063              M2=M1
0064              M1=TEMP
```

```
FORTRAN IV G LEVEL  21                 PEARSN           DATE = 73101          07/06/07             PAGE 0003

0065       120  A1=(M1+1.)*A12/(M1+M2+2.)
0066            A2=A12-A1
0067            YE=(1./A12)*(((M1+1.)**M1*(M2+1.)**M2)/(M1+M2+2.)**(M1+M2))*
               1(GAMMA(M1+M2+2.)/(GAMMA(M1+1.)*GAMMA(M2+1.)))
0068            WRITE (6,2)
0069            X=L1
0070            KK=1
0071       150  IF ( X .GT. L2 ) GO TO 1000
0072            T=X-U1
0073            R1=1.+T/A1
0074            R2=1.-T/A2
0075            IF ( R1 .LE. 0. .AND. R2 .GT. 0. ) P=YE*R2**M2
0076            IF ( R2 .LE. 0. .AND. R1 .GT. 0. ) P=YE*R1**M1
0077            IF ( R1 .LE. 0. .AND. R2 .LE. 0. ) P=YE
0078            IF ( R1 .GT. 0. .AND. R2 .GT. 0. ) P=YE*R1**M1*R2**M2
0079            IF ( R1 .LE. 0. .OR. R2 .LE. 0. ) NHIGH=1
0080            GO TO 5000
          C
          C  CURVE IV *SECOND MAIN TYPE*
          C
0081       200  WRITE (6,5)
0082            R=6.*(B2-B1-1.)/(2.*B2-3.*B1-6.)
0083            M=(R+2.)/2.
0084            A=SQRT((U2/16.)*(16.*(R-1.)-B1*(R-2.)**2))
0085            V=SQRT((R*R*(R-2.)**2*B1)/(16.*(R-1.)-B1*(R-2.)**2))
          C
          C  V AND A HAVE OPPOSITE SIGN OF U3
          C
0086            SGN=1.
0087            IF ( U3 .GT. 0. ) SGN=-1.
0088            A=SGN*ABS(A)
0089            V=SGN*ABS(V)
          C
          C  NOTE ... ARCTAN(-Z)=-ARCTAN(Z)
          C
0090            THETA=ATAN(ABS(V/R))
0091            IF ( V/R .LT. 0. ) THETA=-THETA
0092            COST=COS(ABS(THETA))
          C
          C
0093            IF ( V-2. ) 225,225,250
          C
          C
0094       225  YO=1./(ABS(A)*FRV(R,V))
0095            GO TO 270
          C
          C  APPROXIMATION FOR YO
          C
0096       250  TEMP=(COST      **2)/(3.*R)-1./(12.*R)-THETA*V
0097            YO=(1./A)*SQRT(R/TPI)*EXP(TEMP)/(COST      **(R+1.))
          C
          C
0098       270  CONTINUE
0099            WRITE (6,2)
```

```
FORTRAN IV G LEVEL  21                    PEARSN              DATE = 73101          07/06/07            PAGE 0004

0100             X=L1
0101             Y=V/R
0102             KK=2
0103       280   IF ( X .GT. L2 ) GO TO 1000
0104             T=X-U1
0105             Q=T/A-Y
0106             P=YO*(1.+Q*Q)**(-M)*EXP(-V*ATAN(Q))
0107             GO TO 5000
          C
          C  CURVE VI *THIRD MAIN TYPE*
          C
0108       300   WRITE (6,6)
0109             R=6.*(B2-B1-1.)/(6.+3.*B1-2.*B2)
0110             C=B1*(R+2.)**2+16.*(R+1.)
0111             A=SQRT(U2*C)/2.
          C
          C  A HAS SAME SIGN AS U3
          C
0112             IF ( U3 .LT. 0. ) A=-A
0113             TEMP=R*(R+2.)/2.*SQRT(B1/C)
0114             Q1=-(R-2.)/2.+TEMP
0115             Q2=(R-2.)/2.+TEMP
          C
          C  R IS ALWAYS NEGATIVE HERE.  Q1 IS GREATER THAN Q2
          C
0116             T1=Q2+1.
0117             T2=Q1-Q2-2.
0118             T3=Q1-1.
0119             A1=A*T3/T2
0120             A2=A*T1/T2
0121             YE=(T1**Q2*T2**(Q1-Q2)*GAMMA(Q1))/(ABS(A)
               1  *T3**Q1*GAMMA(T2+1.)*GAMMA(T1))
0122             WRITE (6,2)
0123             X=L1
0124             KK=3
0125       350   IF ( X .GT. L2 ) GO TO 1000
0126             T=X-U1
0127             R1=ABS(1.+T/A1)
0128             R2=ABS(1.+T/A2)
0129             P=YE*R1**(-Q1)*R2**Q2
0130             GO TO 5000
          C
          C  TRANSITION CURVES
          C
0131       400   IF ( Z2 -3. ) 450,410,480
          C
          C  NORMAL CURVE OF ERROR *TRANSITION CURVE*
          C
0132       410   WRITE (6,7)
0133             YO=1./SQRT(TPI*U2)
0134             WRITE (6,2)
0135             X=L1
0136             KK=4
0137       420   IF ( X .GT. L2 ) GO TO 1000
```

```
FORTRAN IV G LEVEL  21              PEARSN          DATE = 73101        07/06/07            PAGE 0005

            P=YO*EXP(-((X-U1)**2)/(2.*U2))
            GO TO 5000
      C
      C  TYPE II CURVE *TRANSITION CURVE*
      C
       450  WRITE (6,8)
            M=(5.*B2-9.)/(6.-2.*B2)
            A2=(2.*U2*B2)/(3.-B2)
            YO=(1./SQRT(A2*PI))*GAMMA(M+1.5)/GAMMA(M+1.)
            WRITE (6,2)
            X=L1
            KK=5
       460  IF ( X .GT. L2 ) GO TO 1000
            T=X-U1
            R1=1.-T*T/A2
            IF ( R1 .LE. 0. ) P=YO
            IF ( R1 .GT. 0. ) P=YO*R1**M
            IF ( R1 .LE. 0. ) NHIGH=1
            GO TO 5000
      C
      C  TYPE VII CURVE *TRANSITION CURVE*
      C
       480  WRITE (6,9)
            M=(5.*B2-9.)/(2.*B2-6.)
            A2=(2.*U2*B2)/(B2-3.)
            YO=(1./SQRT(A2*PI))*GAMMA(M)/GAMMA(M- .5)
            WRITE (6,2)
            X=L1
            KK=6
       490  IF ( X .GT. L2 ) GO TO 1000
            T=X-U1
            P=YO*(1.+T*T/A2)**(-M)
            GO TO 5000
      C
      C  TYPE V CURVE *TRANSITION CURVE*
      C
       500  WRITE (6,10)
            P=4.+(8.+4.*SQRT(4.+B1))/B1
            G=(P-2.)*SQRT(U2*(P-3.))
            YO=G**(P-1.)/GAMMA(P-1.)
            ORIGIN=U1-G/(P-2.)
            IF ( U3 .LT. 0. ) ORIGIN=U1+G/(P-2.)
            WRITE (6,2)
            X=L1
            KK=7
       520  IF ( X .GT. L2 ) GO TO 1000
            X=X-ORIGIN
            T=ABS(X)
            P=YO*T**(-P)*EXP(-G/X)
            GO TO 5000
      C  TYPE III CURVE *TRANSITION CURVE*
      C
       600  IF ( B1 .EQ. 0. ) GO TO 80
            WRITE (6,11)
```

```
FORTRAN IV G LEVEL  21              PEARSN              DATE = 73101          07/06/07              PAGE 0006

0181                LAMBDA=2.*U2/U3
0182                F=4./B1-1.
0183                A=(F+1.)/LAMBDA
0184                YE=LAMBDA*(((F+1.)**F)/(EXP(F+1.)*GAMMA(F+1.)))
0185                WRITE (6,2)
0186                X=L1
0187                KK=8
0188           620  IF ( X .GT. L2 ) GO TO 1000
0189                T=X-U1
0190                R1=1.+T/A
0191                IF ( R1 .LE. 0. ) P=YE*EXP(-LAMBDA*T)
0192                IF ( R1 .GT. 0. ) P=YE*R1**F*EXP(-LAMBDA*T)
0193                IF ( R1 .LE. 0. ) NHIGH=1
0194                GO TO 5000
               C
               C  TYPE VIII CURVE *TRANSITION CURVE*
               C
0195           700  WRITE (6,19)
               C
               C  SOLUTION OF CUBIC EQUATION FOR VALUE OF M
               C
               C  WITH TYPE VIII, M IS RESTRICTED TO 0-1 RANGE
               C
0196                CF(1)=16.*B1
0197                CF(2)=-24.*B1
0198                CF(3)=9.*B1-12.
0199                CF(4)=4.-B1
0200                CALL POLRT (CF,3,RR,RI,IER)
0201                DO 705 I=1,3
0202                M=RR(I)
0203                IF (M.GE.0..AND.M.LE.1..AND.RI(I).EQ.0. ) GO TO 710
0204           705  CONTINUE
0205                WRITE (6,25)
0206                GO TO 750
0207           710  T=(2.-M)*SQRT(U2*(3.-M)/(1.-M))
0208                A=T*(1.-M)/(2.-M)
               C
               C  A HAS OPPOSITE SIGN OF U3
               C
0209                IF ( U3 .LT. 0. ) A=-A
0210                YE=((1.-M)*(2.-M)**M)/(T*(1.-M)**M)
0211                WRITE (6,2)
0212                X=L1
0213                KK=9
0214           720  IF ( X .GT. L2 ) GO TO 1000
0215                T=X-U1
0216                P=YE*(1.-T/A)**(-M)
0217                GO TO 5000
               C
               C  TYPE IX CURVE *TRANSITION CURVE*
               C
0218           750  WRITE (6,20)
               C
               C  SOLUTION OF CUBIC EQUATION FOR VALUE OF M
```

```
      C
      C  WITH TYPE IX, M MUST BE NON-NEGATIVE
      C
             CF(1)=16.*B1
             CF(2)=24.*B1
             CF(3)=9.*B1-12.
             CF(4)=B1-4.
             CALL POLRT (CF,3,RR,RI,IER)
             DO 755 I=1,3
             M=RR(I)
             IF ( M .GE. 0. .AND. RI(I) .EQ. 0. ) GO TO 769
        755  CONTINUE
             WRITE (6,26)
             GO TO 100
        769  T=(M+2.)*SQRT(U2*(M+3.)/(M+1.))
             A=(M+1.)*T/(M+2.)
             YE=(M+1.)**(M+1.)/(T*(M+2.)**M)
             WRITE (6,2)
             X=L1
             KK=10
        770  IF ( X .GT. L2 ) GO TO 1000
             T=X-U1
             P=YE*(1.+T/A)**M
             GO TO 5000
      C
      C  TYPE X CURVE *TRANSITION CURVE*
      C
        800  WRITE (6,21)
             A=SQRT(U2)
             YE=1./(2.718282*A)
             WRITE (6,2)
             X=L1
             KK=11
        820  IF ( X .GT. L2 ) GO TO 1000
             T=(X-U1)
             IF ( U3 .LT. 0. ) P=YE*EXP(T/A)
             IF ( U3 .GE. 0. ) P=YE*EXP(-T/A)
             GO TO 5000
      C
      C  TYPE XI CURVE *TRANSITION CURVE*
      C
        850  WRITE (6,22)
      C
      C  SOLUTION OF CUBIC EQUATION FOR VALUE OF M
      C
      C  WITH TYPE XI, M MUST BE GREATER THAN 1
      C
             CF(1)=16.*B1
             CF(2)=-24.*B1
             CF(3)=9.*B1-12.
             CF(4)=4.-B1
             CALL POLRT (CF,3,RR,RI,IER)
             DO 855 I=1,3
             M=RR(I)
```

```
FORTRAN IV G LEVEL  21                 PEARSN           DATE = 73101          07/06/07             PAGE 0008

            IF ( M .GT. 1. .AND. RI(I) .EQ. 0. ) GO TO 869
      855   CONTINUE
            WRITE (6,24)
            GO TO 300
      869   B=(M-2.)*SQRT(U2*(M-3.)/(M-1.))
            YE=((M-2.)**M)/(B*(M-1.)**(M-1.))
C
C  B HAS SAME SIGN AS U3
C
            B=ABS(B)
            IF ( U3 .LT. 0. ) B=-B
            A=B*(M-1.)/(M-2.)
            WRITE (6,2)
            X=L1
            KK=12
      870   IF ( X .GT. L2 ) GO TO 1000
            T=X-U1
            P=YE*(1.+T/A)**(-M)
            GO TO 5000
C
C  AREA UNDER CURVE APPROXIMATED BY TRAPEZOID FORMULA
C
     1000   AREA=STEP*(CUMP-(X0+XN)/2.)
            WRITE (6,13) AREA
            RETURN
C
     5000   WRITE (6,3) X,P
            IF ( P .GE. 1. ) P=0.
            CUMP=CUMP+P
            IF ( NHIGH .EQ. 0 ) GO TO 5010
            WRITE (6,18)
            NHIGH=0
     5010   IF ( X .EQ. L1 ) X0=P
            IF ( X .EQ. L2 ) XN=P
            X=X+STEP
            GO TO (150,280,350,420,460,490,520,620,720,770,820,870), KK
C
C
        1   FORMAT (   /18H INDEX OF SKEWNESS F21.5/19H DEGREE OF KURTOSIS
           1F20.5/16H CRITERION VALUEF23.5)
        2   FORMAT (5X,5HINDEX,5X,13HRELATIVE FREQ//)
        3   FORMAT (1X,F9.3,F18.5)
        4   FORMAT (/22H TYPE I CURVE SELECTED/)
        5   FORMAT (/23H TYPE IV CURVE SELECTED/)
        6   FORMAT (/23H TYPE VI CURVE SELECTED/)
        7   FORMAT (/28H NORMAL ERROR CURVE SELECTED/)
        8   FORMAT (/23H TYPE II CURVE SELECTED/)
        9   FORMAT (/24H TYPE VII CURVE SELECTED/)
       10   FORMAT (/22H TYPE V CURVE SELECTED/)
       11   FORMAT (/24H TYPE III CURVE SELECTED/)
       13   FORMAT (//15H APPROX AREA ISF15.7//)
       14   FORMAT (////49H APPROXIMATION OF DISTRIBUTION WITH PEARSON CURVE//
           121H ESTIMATED PARAMETERS///5H MEANF34.5/9H VARIANCEF30.5/19H 3RD C
           2ENTRAL MOMENT F20.5/19H 4TH CENTRAL MOMENTF20.5)
```

```
FORTRAN IV G LEVEL  21              PEARSN          DATE = 73101        07/06/07            PAGE 0009

 0301         15    FORMAT (//23H NO CURVE WHEN U(2) = 0//)
 0302         16    FORMAT (//38H INVALID LIMITS AND/OR DELTA ARGUMENTS//)
 0303         18    FORMAT (1H+,30X,32HORDINATE UNDEFINED AT THIS POINT)
 0304         19    FORMAT (/25H TYPE VIII CURVE SELECTED/)
 0305         20    FORMAT (/23H TYPE IX CURVE SELECTED/)
 0306         21    FORMAT (/22H TYPE X CURVE SELECTED/)
 0307         22    FORMAT (/23H TYPE XI CURVE SELECTED/)
 0308         23    FORMAT (/7H LAMBDA,E36.5/10H 5B2-6B1-9,F29.5/10H 2B2-3B1-6,F29.5/)
 0309         24    FORMAT (/22H M .LE. 1. FOR TYPE XI /)
 0310         25    FORMAT (/29H M OUT OF RANGE FOR TYPE VIII /)
 0311         26    FORMAT (/23H M NEGATIVE FOR TYPE IX /)
 0312         27    FORMAT (51H INCORRECT CODE FOR PEARSON CURVE SPECIFIED BY USER/
                   134H CURVE WILL BE SELECTED BY PROGRAM//)
 0313         28    FORMAT (6H CURVEI5,18H REQUESTED BY USER//)
 0314               END
```

```
FORTRAN IV G LEVEL  21                PEARSN          DATE = 73101        07/06/07             PAGE 0010

  *OPTIONS IN EFFECT*  ID,EBCDIC,SOURCE,NOLIST,NODECK,LOAD,NOMAP
  *OPTIONS IN EFFECT*  NAME = PEARSN  , LINECNT =       56
  *STATISTICS*    SOURCE STATEMENTS =      314,PROGRAM SIZE =    10904
  *STATISTICS*  NO DIAGNOSTICS GENERATED
```

```
FORTRAN IV G LEVEL  21              POLRT            DATE = 73101         07/06/07            PAGE 0001

0001              SUBROUTINE POLRT (XCOF,M,ROOTR,ROOTI,IER)
            C  COMPUTATION OF THE REAL AND COMPLEX ROOTS OF A REAL POLYNOMIAL
            C
            C  METHOD:  NEWTON-RAPHSON ITERATIVE TECHNIQUE.  THE FINAL ITERATIONS ON
            C          EACH ROOT ARE PERFORMED USING THE ORIGINAL PCLYNOMIAL RATHER
            C          THAN THE REDUCED POLYNCMIAL TO AVOID ACCUMULATED ERRORS IN THE
            C          POLYNOMIAL.
            C
            C  LIMITATIONS:  LIMITED TO 36TH ORDER POLYNOMIAL OR LESS.
            C                FLOATING POINT OVERFLOW MAY OCCUR FOR HIGH ORDER
            C                POLYNOMIALS BUT WILL NOT AFFECT THE ACCURACY OF THE
            C                RESULTS.
            C
            C  PARAMETERS
            C
            C  XCOF     VECTOR OF M+1 COEFFICIENTS OF THE POLYNOMIAL ORDERED FROM
            C              SMALLEST TO LARGEST POWER
            C  COF      WORKING VECTOR OF LENGTH M+1
            C  M        ORDER OF POLYNOMIAL
            C  ROOTR    RESULTANT VECTOR OF LENGTH M CONTAINING REAL ROOTS
            C  ROOTI    RESULTANT VECTOR OF LENGTH M CONTAINING THE CORRESPONDING
            C            IMAGINARY ROOTS
            C  IER      ERROR CODE RETURNED WHERE
            C              0 = NO ERROR
            C              1 = M LESS THAN 1
            C              2 = M GREATER THAN 36
            C              3 = UNABLE TO DETERMINE ROOT WITH 500 ITERATIONS ON 5
            C                  STARTING VALUES
            C              4 = HIGH ORDER COEFFICIENT IS ZERO
            C
            C  SUBROUTINE TAKEN FROM IBM/SSP
0002              DIMENSION XCOF(1),COF(36),ROOTR(1),ROOTI(1)
0003              IFIT=0
0004              N=M
0005              IER=0
0006              IF ( XCOF(N+1) ) 10,25,10
0007         10   IF ( N ) 15,15,32
            C
            C  SET ERROR CODE TO 1
            C
0008         15   IER=1
0009         20   RETURN
            C  SET ERROR CODE TO 4
            C
0010         25   IER=4
0011              GO TO 20
            C
            C  SET ERROR CODE TO 2
0012         30   IER=2
0013              GO TO 20
0014         32   IF ( N-36 ) 35,35,30
0015         35   NX=N
0016              NXX=N+1
0017              N2=1
```

```
FORTRAN IV G LEVEL  21              POLRT              DATE = 73101          07/06/07          PAGE 0002

 0018                KJ1=N+1
 0019                DO 40 L=1,KJ1
 0020                MT=KJ1-L+1
 0021          40    COF(MT)=XCOF(L)
              C
              C  SET INITIAL VALUES
              C
 0022          45    XO=.5001010E-2
 0023                YO=.1000101E-1
              C
              C  ZERO INITIAL VALUE COUNTER
              C
 0024                IN=0
 0025          50    X=XO
              C
              C  INCREMENT INITIAL VALUES AND COUNTER
              C
 0026                XO=-10.*YO
 0027                YO=-10.*X
              C
              C  SET X AND Y TO CURRENT VALUE
              C
 0028                X=XO
 0029                Y=YO
 0030                IN=IN+1
 0031                GO TO 59
 0032          55    IFIT=1
 0033                XPR=X
 0034                YPR=Y
              C
              C  EVALUATE POLYNOMIAL AND DERIVATIVES
              C
 0035          59    ICT=0
 0036          60    UX=0.
 0037                UY=0.
 0038                V=0.
 0039                YT=0.
 0040                XT=1.
 0041                U=COF(N+1)
 0042                IF ( U ) 65,130,65
 0043          65    DO 70 I=1,N
 0044                L=N-I+1
 0045                XT2=X*XT-Y*YT
 0046                YT2=X*YT+Y*XT
 0047                U=U+COF(L)*XT2
 0048                V=V+COF(L)*YT2
 0049                FI=I
 0050                UX=UX+FI*XT*COF(L)
 0051                UY=UY-FI*YT*COF(L)
 0052                XT=XT2
 0053          70    YT=YT2
 0054                SUMSQ=UX*UX+UY*UY
 0055                IF ( SUMSQ ) 75,110,75
 0056          75    DX=(V*UY-U*UX)/SUMSQ
```

```
FORTRAN IV G LEVEL  21              POLRT             DATE = 73101          07/06/07            PAGE 0003

            X=X+DX
            DY=-(U*UY+V*UX)/SUMSQ
            Y=Y+DY
       78   IF ( ABS(DY)+ABS(DX)-1.0E-5 )  100,80,80
      C
      C  STEP ITERATION COUNTER
      C
       80   ICT=ICT+1
            IF ( ICT-500 ) 60,85,85
       85   IF ( IFIT ) 100,90,100
       90   IF ( IN-5 ) 50,95,95
      C
      C  SET ERROR CODE TO 3
      C
       95   IER=3
            GO TO 20
      100   DO 105 L=1,NXX
            MT=KJ1-L+1
            TEMP=XCOF(MT)
            XCOF(MT)=COF(L)
      105   COF(L)=TEMP
            ITEMP=N
            N=NX
            NX=ITEMP
            IF ( IFIT ) 120,55,120
      110   IF ( IFIT ) 115,50,115
      115   X=XPR
            Y=YPR
      120   IFIT=0
            IF ( X ) 122,125,122
      122   IF ( ABS(Y)-ABS(X)*1.0E-4 ) 135,125,125
      125   ALPHA=X+X
            SUMSQ=X*X+Y*Y
            N=N-2
            GO TO 140
      130   X=0.
            NX=NX-1
            NXX=NXX-1
      135   Y=0.
            SUMSQ=0.
            ALPHA=X
            N=N-1
      140   L1=1
            L2=2
            COF(L2)=COF(L2)+ALPHA*COF(L1)
      145   DO 150 L=2,N
      150   COF(L+1)=COF(L+1)+ ALPHA*COF(L)-SUMSQ*COF(L-1)
      155   ROOTI(N2)=Y
            ROOTR(N2)=X
            N2=N2+1
            IF ( SUMSQ ) 160,165,160
      160   Y=-Y
            SUMSQ=0.
            GO TO 155
```

```
FORTRAN IV G LEVEL  21          POLRT          DATE = 73101          07/06/07          PAGE 0004

      165  IF ( N ) 20,20,45
           END
```

```
FORTRAN IV G LEVEL  21              POLRT           DATE = 73101          07/06/07            PAGE 0005

  *OPTIONS IN EFFECT*  ID,EBCDIC,SOURCE,NOLIST,NODECK,LOAD,NOMAP
  *OPTIONS IN EFFECT*  NAME = POLRT    , LINECNT =       56
  *STATISTICS*    SOURCE STATEMENTS =       106,PROGRAM SIZE =     2350
  *STATISTICS*  NO DIAGNOSTICS GENERATED
```

```
FORTRAN IV G LEVEL  21              RDFMT              DATE = 73101          07/06/07          PAGE 0001

0001                SUBROUTINE RDFMT (FMT, N)
              C
              C  SUBROUTINE FOR INPUTTING VARIABLE FORMAT
              C
              C  INPUT STRUCTURE
              C
              C          FORMAT (ENCLOSED IN PARENTHESES) COLUMNS 01-72
              C             CONTINUE ON CARD 2 IF NECESSARY
              C             CONTINUE ON CARD 3 IF NECESSARY
              C             ...
              C             CONTINUE TO CARD 9 (MAX) IF NECESSARY
              C          "END OF FORMAT" PUNCHED IN COLUMNS 01-13
              C
0002                DIMENSION FMT(162)
0003                DATA END,BLK/3HEND,1H /
0004                DO 50 I=1,162
0005           50   FMT(I)=BLK
0006                N=1
0007                DO 100 I=1,10
0008                M=N+17
0009                READ (5,1)  (FMT(J),J=N,M)
0010                IF ( FMT(N) .EQ. END ) RETURN
0011          100   N=N+18
0012                WRITE (6,2)
0013                STOP
0014            1   FORMAT (18A4)
0015            2   FORMAT (37H *** EXCESSIVE NUMBER OF FORMAT CARDS)
0016                END
```

```
FORTRAN IV G LEVEL  21                RDFMT            DATE = 73101          07/06/07            PAGE 0002

  *OPTIONS IN EFFECT*  ID,EBCDIC,SOURCE,NOLIST,NODECK,LOAD,NOMAP
  *OPTIONS IN EFFECT*  NAME = RDFMT    , LINECNT =       56
  *STATISTICS*    SOURCE STATEMENTS =       16,PROGRAM SIZE =      650
  *STATISTICS*  NO DIAGNOSTICS GENERATED
```

```
FORTRAN IV G LEVEL  21          SETUP          DATE = 73101          07/06/07          PAGE 0001

          SUBROUTINE SETUP (NI,NJ,NK,XMULT)
C
C  COMPUTATION OF BOUNIARIES FOR ESTIMATED CENTRAL MOMENTS:
C  PREPARATION OF X MATRIX FOR USE BY LP SUBROUTINE
C
C  SUBROUTINE WRITTEN ORIGINALLY BY DR. RICHARD K. HILL,JR
C  VIRGINIA POLYTECHNIC INSTITUTE
C
          COMMON /BLOCK5/ SSUM(4),X(10,1007),B(6,4),W(6)
          DIMENSION G(10),OBJF(4)
          DATA G/1H0,1H1,1H2,1H3,1H4,1H5,1H6,1H7,1H8,1H9/
          DATA OBJF/4HOBJE,4HCT F,4HUNCT,4HION /
          DATA EQU,F,BLANK/3HEQU,1HF,1H /
          NI1=NI+1
          NI2=NI+2
          NI3=NI+3
          NI4=NI+4
          X(1,5)=NI
          DO 82 J=7,NJ
          X(NI1,J)=F
          JH=(J-7)/100
          JT=(J-7-JH*100)/10
          JU=J-7-JH*100-JT*10
          X(NI2,J)=G(JH+1)
          X(NI3,J)=G(JT+1)
    82    X(NI4,J)=G(JU+1)
          DO 83 I=3,NI
          X(I,1)=EQU
          X(I,2)=BLANK
          JT=(I-3)/10
          JU=I-3-JT*10
          X(I,3)=G(JT+1)
    83    X(I,4)=G(JU+1)
          DO 84 I=4,NI
          DO 84 J=7,NJ
          XJ=J
    84    X(I,J)=(XJ-7.)**(I-3)
          DO 85 J=7,NJ
          X(3,J)=1.
          XJ=J
    85    X(1,J)=((XJ-7.)**(NK -2))*XMULT
          DO 86 I=4,NI
    86    X(I,5)=SSUM(I-3)
          X(3,5)=1.
          DO 93 J=1,4
    93    X(2,J)=OBJF(J)
          RETURN
          END
```

```
FORTRAN IV G LEVEL  21                 SETUP              DATE = 73101           07/06/07            PAGE 0002

  *OPTIONS IN EFFECT*  ID,EBCDIC,SOURCE,NOLIST,NODECK,LOAD,NOMAP
  *OPTIONS IN EFFECT*  NAME = SETUP    , LINECNT =        56
  *STATISTICS*    SOURCE STATEMENTS =        41,PROGRAM SIZE =      1388
  *STATISTICS*  NO DIAGNOSTICS GENERATED
```

```
FORTRAN IV G LEVEL  21                MOMVAL             DATE = 73101          07/06/07                PAGE 0001
              SUBROUTINE MOMVAL (VAL,MOM,VALC)
C
C  CALCULATION OF BOUNDARIES FOR ESTIMATED CENTRAL MOMENTS:
C  CONVERSION OF FREQUENCIES IN X MATRIX FROM ALPHA TO NUMERIC AND
C  CALCULATION OF MOMENT
C
C  SUBROUTINE WRITTEN ORIGINALLY BY DR. RICHARD K. HILL, JR
C  (VIRGINIA POLYTECHNIC INSTITUTE) AND MODIFIED BY DAVID M. SHOEMAKER
C
      COMMON /BLOCK5/ SSUM(4),X(10,1007),B(6,4),W(6)
      DIMENSION G(10)
      DATA G/1H0,1H1,1H2,1H3,1H4,1H5,1H6,1H7,1H8,1H9/
      NI=X(1,5)
      VAL=0.
      DO 97 I=3,NI
      DO 98 L=1,10
      IF ( X(I,2) .EQ. G(L) ) GO TO 99
   98 CONTINUE
   99 SAVE=(L-1)*100
      DO 100 L=1,10
      IF ( X(I,3) .EQ. G(L) ) GO TO 101
  100 CONTINUE
  101 SAVE=SAVE+(L-1)*10
      DO 102 L=1,10
      IF ( X(I,4) .EQ. G(L) ) GO TO 103
  102 CONTINUE
  103 SAVE=SAVE+L-1
   97 VAL=VAL+X(I,6)*(SAVE**MOM)
      V1=SSUM(1)
      V2=SSUM(2)
      V3=SSUM(3)
      K=MOM-1
      GO TO (200,300,400), K
  200 VALC=VAL-V1*V1
      RETURN
  300 VALC=VAL-3.*V1*V2+2.*V1**3
      RETURN
  400 VALC=VAL-4.*V3*V1+6.*V2*V1*V1-3.*V1**4
      RETURN
      END
```

```
FORTRAN IV G LEVEL  21                MOMVAL           DATE = 73101          07/06/07           PAGE 0002

 *OPTIONS IN EFFECT*  ID,EBCDIC,SOURCE,NOLIST,NODECK,LOAD,NOMAP
 *OPTIONS IN EFFECT*  NAME = MOMVAL  , LINECNT =       56
 *STATISTICS*    SOURCE STATEMENTS =       32,PROGRAM SIZE =     1136
 *STATISTICS*  NC DIAGNOSTICS GENERATED
```

```
FORTRAN IV G LEVEL  21              SUBTST            DATE = 73101        07/06/07              PAGE 0001

0001                SUBROUTINE SUBTST (KPOP,NDISK)
             C
             C   CALCULATION OF PARAMETER ESTIMATES FROM SUBTEST RESULTS
             C
0002                IMPLICIT REAL*8 (A-H,O-Z)
0003                REAL*4 XX,FMT,RSLT,SDATA
0004                INTEGER*4 FS,AS,CAS,CFS
0005                COMMCN /BLOCK1/ SDATA(102,11),NTEST
0006                COMMCN  /BLOCK8/ FS(561),AS(55),CAS(55),CFS(561)
0007                DIMENSION XIP(100),YIP(100),ZIP(100),UIP(100),VIP(100),
                   2          XPJ(150),YPJ(150),ZPJ(150),UPJ(150),VPJ(150),
                   3          XX(100,150),S2(4),S3(10),S4(33),C2(4),C3(10),C4(33),
                   4          D(4),A(10),F(33),DD(4),AA(10),FF(33),FMT(162),RSLT(8)
             C
             C
0008                NRDR=5
0009                IF ( NDISK .GT. 0 ) NRDR=2
             C
             C
0010                NGOOD=0
0011                SEU1=0.
0012                SEU2=0.
0013                SVCIT=0.
0014                SP2=0.
             C
0015                WRITE (6,3)
0016                DO 500 IJ=1,NTEST
             C
0017                READ (5,7) NSPT,IPT,NCARD
0018                IF ( IPT .GT. 0 ) GO TO 40
             C
             C   WHEN THE NUMBER OF ITEMS PER SUBTEST IS SET EQUAL TC ZERO, THE ITEM
             C   SCORES FROM THAT PARTICULAR SUBTEST ARE IGNORED COMPLETELY.
             C
0019                NSKIP=NSPT*NCARD
0020                DO 30 J=1,NSKIP
0021          30    READ (NRDR,9) NFAN
0022                DO 35 J=1,8
0023          35    SDATA(IJ,J)=-9999999.
0024                SDATA(IJ,10)=0.
0025                WRITE (6,5) IJ
0026                GO TO 500
0027          40    CALL RDFMT (FMT,NC)
0028                NC=NC-1
             C
             C   IF THERE ARE MORE THAN 100 SUBTESTS, THE REMAINING SUBTEST RESULTS
             C   ARE IGNORED.  ALSO, ANY SUBTEST HAVING MORE THAN 150 ITEMS IS
             C   IGNORED.
             C
0029                IF ( IJ .LE. 100 .AND. IPT .LE. 150 ) GO TO 50
0030                WRITE (6,5) IJ
0031                NSKIP=NSPT*NCARD
0032                DO 45 J=1,NSKIP
0033          45    READ(NRDR,9) NFAN
```

```
FORTRAN IV G LEVEL  21            SUBTST         DATE = 73101        07/06/07          PAGE 0002

              GO TO 500
       50     CONTINUE
      C
      C
              SDATA(IJ,9)=NSPT
              SDATA(IJ,10)=IPT
              WRITE (6,4)IJ,NSPT,IPT,(FMT(J),J=1,NC)
      C
      C  LOAD ITEM SCORES INTO CORE MATRIX
      C
              DO 55 I=1,NSPT
       55     READ (NRDR,FMT) (XX(I,J),J=1,IPT)
      C
      C  IF THE NUMBER OF ITEMS FOR A PARTICULAR SUBTEST IS LESS THAN FOUR,
      C  ONLY THE MEAN TEST SCORE IS ESTIMATED.
      C
              IF ( IPT .GT. 3 ) GO TO 75
              XPP=0.
              DO 65 J=1,IPT
              XPJ(J)=0.
              DO 60 I=1,NSPT
       60     XPJ(J)=XPJ(J)+XX(I,J)
              XPP=XPP+XPJ(J)
       65     XPJ(J)=XPJ(J)/NSPT
              EU1=(XPP/(NSPT*IPT))*KPOP
              WRITE (6,6)  (XPJ(J),J=1,IPT)
              SDATA(IJ,1)=EU1
              DO 70 J=2,8
       70     SDATA(IJ,J)=-9999999.
              GO TO 500
      C
      C
       75     CONTINUE
      C
      C
              CALL VCOV (XX,NSPT,IPT,VCOVIT)
              SVCIT=SVCIT+VCOVIT
      C
      C
              DO 80 I=1,IPT
              XPJ(I)=0.
              YPJ(I)=0.
              ZPJ(I)=0.
              UPJ(I)=0.
       80     VPJ(I)=0.
              DO 85 I=1,33
       85     S4(I)=0.
      C
      C
              DO 100 I=1,NSPT
              XIP(I)=0.
              YIP(I)=0.
              ZIP(I)=0.
              UIP(I)=0.
```

```
FORTRAN IV G LEVEL  21          SUBTST          DATE = 73101          07/06/07          PAGE 0003

            VIP(I)=0.
            DO 100 J=1,IPT
            TEMP=XX(I,J)
            XIP(I)=XIP(I)+TEMP
            XPJ(J)=XPJ(J)+TEMP
            YIP(I)=YIP(I)+TEMP**2
            YPJ(J)=YPJ(J)+TEMP**2
            ZIP(I)=ZIP(I)+TEMP**3
      100   ZPJ(J)=ZPJ(J)+TEMP**3
            Q2=0.
            Q4=0.
            Q7=0.
            DO 120 I=1,IPT
            TEMP=XPJ(I)
            Q2=Q2+TEMP**2
            Q4=Q4+TEMP**3
            Q7=Q7+YPJ(I)*XPJ(I)
            S4(9)=S4(9)+TEMP**4
            S4(30)=S4(30)+YPJ(I)**2
      120   S4(32)=S4(32)+XPJ(I)*ZPJ(I)
            Q1=0.
            Q3=0.
            Q6=0.
            DO 130 I=1,NSPT
            TEMP=XIP(I)
            Q1=Q1+TEMP**2
            Q3=Q3+TEMP**3
            Q6=Q6+YIP(I)*XIP(I)
            S4(8)=S4(8)+TEMP**4
            S4(29)=S4(29)+YIP(I)**2
            S4(31)=S4(31)+XIP(I)*ZIP(I)
      130   S4(22)=S4(22)+YIP(I)*TEMP**2
            Q8=0.
            S3(6)=0.
            S3(10)=0.
            DO 145 I=1,NSPT
            DO 140 J=1,IPT
            TEMP=XX(I,J)*XIP(I)
            UIP(I)=UIP(I)+TEMP
            UPJ(J)=UPJ(J)+TEMP
            TEMP=XX(I,J)*XPJ(J)
            VIP(I)=VIP(I)+TEMP
            VPJ(J)=VPJ(J)+TEMP
            Q8=Q8+XX(I,J)**4
            S3(10)=S3(10)+XX(I,J)**3
            TEMP=XX(I,J)*XIP(I)*XPJ(J)
            S3(6)=S3(6)+TEMP
      140   S4(26)=S4(26)+XX(I,J)*TEMP
            S4(14)=S4(14)+VIP(I)**2
            S4(24)=S4(24)+YIP(I)*VIP(I)
      145   S4(15)=S4(15)+UIP(I)*VIP(I)
            Q5=0.
            XPP=0.
            YPP=0.
```

```
FORTRAN IV G LEVEL  21                  SUBTST            DATE = 73101         07/06/07              PAGE 0004

0125                ZPP=0.
0126                DO 150 I=1,IPT
0127                XPP=XPP+XPJ(I)
0128                YPP=YPP+YPJ(I)
0129                ZPP=ZPP+ZPJ(I)
0130                Q5=Q5+UPJ(I)*XPJ(I)
0131                S4(13)=S4(13)+UPJ(I)**2
0132                S4(16)=S4(16)+UPJ(I)*VPJ(I)
0133                S4(23)=S4(23)+YPJ(I)*XPJ(I)**2
0134          150   S4(25)=S4(25)+YPJ(I)*UPJ(I)
0135                Q9=0.
0136                NSTOP=IPT-1
0137                DO 160 I=1,NSTOP
0138                NSTART=I+1
0139                DO 160 J=NSTART,IPT
0140                TEMP=0.
0141                DO 155 K=1,NSPT
0142          155   TEMP=TEMP+XX(K,I)*XX(K,J)
0143          160   Q9=Q9+2.*TEMP*TEMP
0144                DO 165 I=1,IPT
0145          165   Q9=Q9+YPJ(I)**2
              C
              C  CALCULATION OF SIGMAS REQUIRED IN FIRST, SECOND, THIRD AND FOURTH
              C  DEGREE GSM'S
              C
0146                S4(1)=XPP**4
0147                S4(2)=XPP*XPP*Q1
0148                S4(3)=XPP*XPP*Q2
0149                S4(4)=Q1*Q1
0150                S4(5)=Q2*Q2
0151                S4(6)=XPP*Q3
0152                S4(7)=XPP*Q4
0153                S4(10)=XPP*Q5
0154                S4(11)=Q1*Q2
0155                S4(12)=XPP*XPP*YPP
0156                S4(17)=Q9
0157                S4(18)=YPP*Q1
0158                S4(19)=YPP*Q2
0159                S4(20)=XPP*Q6
0160                S4(21)=XPP*Q7
0161                S4(27)=YPP*YPP
0162                S4(28)=XPP*ZPP
0163                S4(33)=Q8
              C
              C
0164                S3(1)=XPP**3
0165                S3(2)=XPP*Q1
0166                S3(3)=XPP*Q2
0167                S3(4)=Q3
0168                S3(5)=Q4
0169                S3(7)=YPP*XPP
0170                S3(8)=Q6
0171                S3(9)=Q7
              C
```

```
FORTRAN IV G LEVEL  21              SUBTST          DATE = 73101        07/06/07          PAGE 0005
      C
            S2(1)=XPP**2
            S2(2)=Q1
            S2(3)=Q2
            S2(4)=YPP
      C
      C
            S1=XPP
      C
      C  CALCULATION OF CONSTANTS ASSOCIATED WITH GSM'S
      C
            RC=NSPT*IPT
            R1=NSPT-1.
            R2=NSPT-2.
            R3=NSPT-3.
            A1=IPT-1.
            A2=IPT-2.
            A3=IPT-3.
      C
      C  CONSTANT FOR FIRST LEGREE GSM
      C
            C1=1./RC
      C
      C  CONSTANTS FOR SECOND DEGREE GSM'S
      C
            C2(1)=1./(RC*R1*A1)
            C2(2)=1./(RC*A1)
            C2(3)=1./(RC*R1)
            C2(4)=1./RC
      C
      C
            T=RC*R1*A1
            U=1./(T*R2*R3*A2*A3)
            B=1./(T*R2*A2*A3)
            C=1./(T*R2*R3*A2)
            V=1./(T*A2*A3)
            E=1./(T*R2*R3)
            G=1./(T*R2*A2)
            H=1./(T*A2)
            P=1./(T*R2)
            Q=1./T
      C
      C  CONSTANTS FOR THIRD DEGREE GMS'S
      C
            C3(1)=G
            C3(2)=H
            C3(3)=P
            C3(4)=1./(RC*A1*A2)
            C3(5)=1./(RC*R1*R2)
            C3(6)=Q
            C3(7)=Q
            C3(8)=1./(RC*A1)
            C3(9)=1./(RC*R1)
            C3(10)=1./RC
```

```
FORTRAN IV G LEVEL  21            SUBTST          DATE = 73101          07/06/07          PAGE 0006

      C
      C  CONSTANTS FOR FOURTH DEGREE GSM'S
      C
            C4(1)=U
            C4(2)=B
            C4(3)=C
            C4(4)=V
            C4(5)=E
            C4(6)=V
            C4(7)=E
            C4(8)=1./(RC*A1*A2*A3)
            C4(9)=1./(RC*R1*R2*R3)
            C4(10)=G
            C4(11)=G
            C4(12)=G
            C4(13)=H
            C4(14)=P
            C4(15)=H
            C4(16)=P
            C4(17)=Q
            C4(18)=H
            C4(19)=P
            C4(20)=H
            C4(21)=P
            C4(22)=1./(RC*A1*A2)
            C4(23)=1./(RC*R1*R2)
            DO 180 I=24,28
        180 C4(I)=Q
            C4(29)=1./(RC*A1)
            C4(30)=1./(RC*R1)
            C4(31)=1./(RC*A1)
            C4(32)=1./(RC*R1)
            C4(33)=1./RC
      C
      C  CALCULATION OF FIRST DEGREE GSM (FIRST DEGREE BIPOLYKAY)
      C
            Z1=S1*C1
      C
      C  CALCULATION OF SECOND DEGREE GSM'S
      C
            D(1)=(S2(1)-S2(2)-S2(3)+S2(4))*C2(1)
            D(2)=(S2(2)-S2(4))*C2(2)
            D(3)=(S2(3)-S2(4))*C2(3)
            D(4)=S2(4)*C2(4)
      C
      C  CALCULATION OF SECOND DEGREE BIPOLYAKYS
      C
            DD(1)=D(1)
            DD(2)=-D(1)+D(2)
            DD(3)=-D(1)+D(3)
            DD(4)=D(1)-D(2)-D(3)+D(4)
      C
      C  CALCULATION OF THIRD DEGREE GSM'S
      C
```

```
FORTRAN IV G LEVEL  21            SUBTST            DATE = 73101         07/06/07            PAGE 0007

0248              DO 200 J=1,10
0249              A(J)=0.
0250              DO 210 I=J,10
0251              LSUB=(J-1)*10+I-J*(J-1)/2
0252        210   A(J)=A(J)+AS(LSUB)*S3(I)
0253        200   A(J)=A(J)*C3(J)
            C
            C  CALCULATION OF THIRD DEGREE BIPOLYKAYS
            C
0254              DO 220 J=1,10
0255              AA(J)=0.
0256              DO 220 I=1,J
0257              LSUB=(J-1)*10+I-(-20+21*J-J*J)/2
0258        220   AA(J)=AA(J)+CAS(LSUB)*A(I)
            C
            C  CALCULATION OF FOURTH DEGREE GSM'S
            C
0259              DO 230 J=1,33
0260              F(J)=0.
0261              DO 240 I=J,33
0262              LSUB=(J-1)*33+I-J*(J-1)/2
0263        240   F(J)=F(J)+FS(LSUB)*S4(I)
0264        230   F(J)=F(J)*C4(J)
            C  CALCULATION OF FOURTH DEGREE BIPOLYKAYS
            C
0265              DO 245 J=1,33
0266              FF(J)=0.
0267              DO 245 I=1,J
0268              LSUB=(J-1)*33+I-(-65+67*J-J*J)/2
0269        245   FF(J)=FF(J)+CFS(LSUB)*F(I)
            C
            C
            C  ESTIMATION OF CENTRAL MOMENTS OF UNIVERSE SCORES WHEN POPULATION
            C  IS INFINITE
            C
            C  MEAN UNIVERSE SCORE
            C
0270              EU1=Z1*KPOP
            C
            C  SECOND CENTRAL MOMENT
            C
0271              EU2=DD(1)*KPOP+DD(2)*KPOP**2
            C
            C  THIRD CENTRAL MOMENT
            C
0272              EU3=AA(4)*KPOP**3+3.*AA(8)*KPOP**2+KPOP*AA(10)
            C
            C  FOURTH CENTRAL MOMENT
            C
0273              EU4=KPOP**4*(3.*FF(4)+FF(8))+KPOP**3*(12.*FF(13)+6.*FF(18)
                 1    +2.*FF(22))+KPOP**2*(6.*FF(17)+12.*FF(25)+3.*FF(27)
                 2    +4.*FF(31))+KPOP*(3.*FF(30)+FF(33))
            C
            C  CALCULATION OF COMPONENTS OF VARIANCE FOR SUBTEST MATRIX AND
```

```
FORTRAN IV G LEVEL  21              SUBTST          DATE = 73101        07/06/07            PAGE 0008

              C  THETA (A FUNCTION OF THE MEAN ITEM RELIABILITY)
              C
 0274               CXX=XPP*XPP*C2(4)
 0275               SSE=Q1/IPT-CXX
 0276               SSI=Q2/NSPT-CXX
 0277               SSIE=XPP-CXX-SSE-SSI
 0278               ERR=SSIE/(A1*R1)
 0279               RSLT(5)=(SSI/A1-ERR)/NSPT
 0280               RSLT(6)=(SSE/R1-ERR)/IPT
 0281               RSLT(7)=ERR
 0282               FMP=(R1*A1-2.)/(R1*A1)
 0283               RSLT(8)=(SSE/R1-FMP*ERR)/(IPT*FMP*ERR)
              C
              C
 0284               RSLT(1)=EU1
 0285               RSLT(2)=EU2
 0286               RSLT(3)=EU3
 0287               RSLT(4)=EU4
              C
              C  ESTIMATION OF VARIANCE OF SECOND CENTRAL MOMENT
              C
 0288               VARU2=(1./NSPT)*(FF(8)+2.*FF(22)/KPOP+FF(29)/KPOP**2)+
                   1 (1./IPT-1./KPOP)*(4.*FF(13)+4.*FF(25)/KPOP+FF(30)/KPOP**2)+
                   2 (2./(NSPT-1.))*(FF(4)+2.*FF(13)/KPOP+FF(17)/KPOP**2)
                   3 +(2./KPOP**2)*(1./(IPT-1.)-1./(KPOP-1.))*FF(17)+(1.-2./KPOP)*
                   4 (2./(IPT*(IPT-1.))-2./(KPOP*(KPOP-1.)))*FF(17)+(2./(IPT*(NSPT-1.)
                   5  ))*(2.*(FF(13)+FF(18))+4.*FF(25)/KPOP+FF(30)/KPOP**2)+(2./(KPOP
                   6  **2*NSPT*(IPT-1.)))*FF(29)+(1./(NSPT*IPT))*(4.*FF(22)+4.*FF(31)/
                   7 KPOP+FF(33)/KPOP**2)+(2./(KPOP**2*(NSPT-1.)*(IPT-1.)))*(FF(17)+
                   8 FF(27))+(1.-2./KPOP)*(2./(IPT*NSPT*(IPT-1.)))*FF(29)+(1.-2./KPOP)
                   9 *(2./(IPT*(NSPT-1.)*(IPT-1.)))*(FF(17)+FF(27))
 0289               VARU2 =VARU2 *KPOP**4
              C
              C  ESTIMATION OF VARIANCE OF MEAN
              C
              C
 0290               VARU1=DD(4)*C1+DD(2)/NSPT+(1./IPT-1./KPOP)*DD(3)
 0291               VARU1=VARU1*KPOP**2
              C
              C
 0292               DO 250 I=1,8
 0293          250  SDATA(IJ,I)=RSLT(I)
              C
              C
 0294               SEU1=SEU1+VARU1
 0295               SEU2=SEU2+VARU2
 0296               SP2=SP2+RSLT(5)
 0297               NGOOD=NGOOD+1
              C
 0298               DO 260 I=1,IPT
 0299          260  XPJ(I)=XPJ(I)/NSPT
 0300               WRITE (6,6)  (XPJ(I),I=1,IPT)
              C
 0301          500  CONTINUE
```

```
FORTRAN IV G LEVEL  21              SUBTST          DATE = 73101        07/06/07          PAGE 0009

                C
                C  COMPUTATION OF STANDARD ERROR OF POOLED ESTIMATE OF MEAN
                C  AND VARIANCE
                C
 0302                 SP2=SP2/NGOOD
 0303                 SAVE=SEU1
 0304                 VARU1=SEU1/NGOOD**2-(KPOP*KPOP*(NGOOD-1.)*SP2)/(NGOOD*(KPOP-1.))
 0305                 IF ( NT*IPT .GT. KPOP ) VARU1=SAVE/NGOOD**2
 0306                 SEU1=0.
 0307                 IF ( VARU1 .GT. 0. ) SEU1=DSQRT(VARU1)
 0308                 VARU2=SEU2/NGOOD**2-(4.*(NGOOD-1.)*SVCIT*KPOP)/(NGOOD*NGOOD)
 0309                 SEU2=0.
 0310                 IF ( VARU2 .GT. 0. ) SEU2=DSQRT(VARU2)
                C
 0311                 IF ( NTEST .GT. 100 ) NTEST=100
                C
                C
 0312                 WRITE (6,1)
 0313                 DO 550 I=1,NTEST
 0314                 IF ( SDATA(I,1) .NE. -9999999. ) GO TO 505
 0315                 WRITE (6,2) I
 0316                 GO TO 550
 0317            505  NN=SDATA(I,10)
 0318                 IF ( NN .LT. 4 ) GO TO 510
 0319                 WRITE (6,2) I,(SDATA(I,J),J=1,8)
 0320                 GO TO 550
 0321            510  WRITE (6,2) I,SDATA(I,1)
 0322            550  CONTINUE
 0323                 IF ( NGOOD .GT. 0 ) WRITE (6,10) SEU1,SEU2
                C
                C
 0324                 RETURN
                C
                C
 0325             1   FORMAT (///25X,21HESTIMATES OF MOMENTS ,33X,
                    122HCOMPONENTS OF VARIANCE/8X,
                    257H---------------------------------------------------------,3X
                    3,57H---------------------------------------------------------/
                    48H SUBTEST,5X,4HMEAN, 9X,8HVARIANCE,10X,2HU3,13X,2HU4,
                    511X,6HVAR(I), 9X,6HVAR(E), 8X,7HVAR(IE), 9X,6HTHETA'//)
 0326             2   FORMAT (1X,I4,8E15.5)
 0327             3   FORMAT (//8H SUBTEST,2X,9HEXAMINEES,3X,5HITEMS,3X,6HFORMAT//)
 0328             4   FORMAT (/1X,I4,2I10,5X,25A4/(30X,25A4))
 0329             5   FORMAT (8H SUBTESTI5,8H IGNORED)
 0330             6   FORMAT (/30X,23HITEM DIFFICULTY INDICES//(30X,10F8.3))
 0331             7   FORMAT (3I4)
 0332             9   FORMAT (A4)
 0333            10   FORMAT (//5X,14HSE(MEAN)-HLN =F10.5/5X,14HSE(VAR )-HLN =F10.5)
 0334                 END
```

```
FORTRAN IV G LEVEL  21                  SUBTST           DATE = 73101          07/06/07            PAGE 0010

 *OPTIONS IN EFFECT*  ID,EBCDIC,SOURCE,NOLIST,NODECK,LOAD,NOMAP
 *OPTIONS IN EFFECT*  NAME = SUBTST  , LINECNT =       56
 *STATISTICS*    SOURCE STATEMENTS =      334,PROGRAM SIZE =    84148
 *STATISTICS*  NO DIAGNOSTICS GENERATED
```

```
FORTRAN IV G LEVEL  21                TTABLE              DATE = 73101          07/06/07                PAGE 0001

0001                 SUBROUTINE TTABLE (T,N,AREA)
               C
               C  CALCULATION OF STUDENT'S T CORRESPONDING TO PRESCRIBED ALPHA LEVEL
               C  (TWO-TAILED PROBABILITY)
               C
               C     ARGUMENTS
               C
               C        T = STUDENT'S T
               C        N = DEGREES OF FREEDOM ASSOCIATED WITH T
               C     AREA = RELATIVE AREA ASSOCIATED WITH CONFIDENCE INTERVAL
               C
               C  POLYNOMIAL APPROXIMATIONS USED FOR BOTH T AND Z
               C
               C  REFERENCE:  ABRAMOWITZ, M. & STEGUM, I. A. HANDBOOK OF MATHEMATICAL
               C              FUNCTIONS. NATIONAL BUREAU OF STANDARDS APPLIED
               C              MATHEMATICS SERIES NO. 55, 1964.
               C              (SEE 26.2.23 AND 26.7.5)
               C
0002                 IF ( N .GT. 0 ) GO TO 100
0003                 WRITE (6,1) N
0004                 GO TO 150
0005             100 P=(1.-AREA)/2.
0006                 IF ( P .GT. 0. .AND. P .LE. .5 ) GO TO 200
0007                 WRITE (6,2) P
0008             150 T=0.
0009                 RETURN
0010             200 X=SQRT(ALOG(1./(P*P)))
               C
0011                 Z=X-(2.515517+.802853*X+.010328*X*X)/
                    1     (1.+1.432788*X+.189269*X*X+.001308*X**3)
               C
0012                 Z3=Z**3
0013                 Z5=Z**5
0014                 Z7=Z**7
0015                 Z9=Z**9
0016                 G1=(Z3+Z)/4.
0017                 G2=(5.*Z5+16.*Z3+3.*Z)/96.
0018                 G3=(3.*Z7+19.*Z5+17.*Z3-15.*Z)/384.
0019                 G4=(79.*Z9+776.*Z7+1482.*Z5-1920.*Z3-945.*Z)/92160.
               C
0020                 T=Z+G1/N+G2/N**2+G3/N**3+G4/N**4
               C
0021                 RETURN
0022               1 FORMAT (//27H ERROR IN T-TABLE ARG: DF =I12//)
0023               2 FORMAT (//26H ERROR IN T-TABLE ARG: P =E15.5//)
0024                 END
```

```
FORTRAN IV G LEVEL  21                TTABLE          DATE = 73101          07/06/07          PAGE 0002

 *OPTIONS IN EFFECT*  ID,EBCDIC,SOURCE,NOLIST,NODECK,LOAD,NOMAP
 *OPTIONS IN EFFECT*  NAME = TTABLE  , LINECNT =       56
 *STATISTICS*    SOURCE STATEMENTS =       24,PROGRAM SIZE =     1320
 *STATISTICS*  NO DIAGNOSTICS GENERATED
```

```
C
C  COMPUTATION OF VARIANCE OF ITEM-TOTAL COVARIANCES FOR ONE SUBTEST
C
      DIMENSION XX(100,150),T(150)
      SUMT=0.
      DO 100 I=1,NSPT
      T(I)=0.
      DO 150 J=1,IPT
  150 T(I)=T(I)+XX(I,J)
      T(I)=T(I)/IPT
  100 SUMT=SUMT+T(I)
      SUMC=0.
      SUMC2=0.
      DO 200 I=1,IPT
      COV=0.
      TEMP=0.
      DO 250 J=1,NSPT
      TEMP=TEMP+XX(J,I)
  250 COV=COV+XX(J,I)*T(J)
      COV=COV/NSPT-TEMP*SUMT/NSPT**2
      SUMC=SUMC+COV
  200 SUMC2=SUMC2+COV*COV
      VCOVIT=SUMC2/IPT-SUMC*SUMC/IPT**2
      RETURN
      END
```

```
FORTRAN IV G LEVEL  21                 VCOV              DATE = 73101          07/06/07              PAGE 0002

  *OPTIONS IN EFFECT*  ID,EBCDIC,SOURCE,NOLIST,NODECK,LOAD,NOMAP
  *OPTIONS IN EFFECT*  NAME = VCOV    , LINECNT =        56
  *STATISTICS*    SOURCE STATEMENTS =       23,PROGRAM SIZE =     1502
  *STATISTICS*  NO DIAGNOSTICS GENERATED

  *STATISTICS*  NO DIAGNOSTICS THIS STEP 2
```

```
IEF285I   SYS1.FORTRAN                                    PASSED
IEF285I   VOL SER NOS= SYSTM3.
IEF285I   SYS73101.T070543.RV000.AGB005Z1.LOADSET         PASSED
IEF285I   VCL SEF NOS= SYSDA1.
IEF285I   SYS73101.T070543.SV000.AGB005Z1.R0000001        SYSOUT
IEF285I   VOL SER NOS= SYSDA4.
IEF285I   SYS73101.T070543.RV000.AGB005Z1.S0000002        SYSIN
IEF285I   VOL SER NOS= SYSDA2.
IEF285I   SYS73101.T070543.RV000.AGB005Z1.S0000002        DELETED
IEF285I   VOL SER NOS= SYSDA2.
CCN011I STEP RETURN CODE =    0

CCN301I   SYSPRINT PRINTED    68 TOTAL PAGES, REQUIRED    40 TRACKS.

  REGION ALLOCATION        CORE USED          MAIN STORAGE       EXTENDED STORAGE
        300K                  158K                300K                 0K

   STEP CPU TIME          JOB CPU TIME       STEP I/O COUNT      JOB I/O COUNT
       15.89S                15.89S               350                 350

XXLKED    EXEC  PGM=IEWL,REGION=&RL,TIME=&TL,COND=(5,LT,FORT),PARM=MAP   00000090
IEF653I SUBSTITUTION JCL - PGM=IEWL,REGION=154K,TIME=1439,COND=(5,LT,FORT),PARM=MAP
XXSYSLIB DD DISP=(SHR,PASS),DSN=$YS&LEVEL..FORTLIB                        00000100
IEF653I SUBSTITUTION JCL - DISP=(SHR,PASS),DSN=$YS1.FORTLIB
XXSYSLIN DD DSN=&&LOADSET,DISP=(OLD,DELETE)                               00000110
XX       DD    DDNAME=SYSIN                                               00000120
XXSYSLMOD DD DISP=(MOD,PASS),DSN=&&GOSET(GOFORTG),UNIT=SYSDA,             00000130
XX SPACE=(3072,(20,20,1),RLSE)                                            00000140
XXSYSUT1 DD UNIT=SYSDA,SPACE=(TRK,(20,20))                                00000150
XXSYSPRINT DD SYSOUT=A,SPACE=(TRK,&PL,RLSE),DCB=BLKSIZE=2178              00000160
IEF653I SUBSTITUTION JCL - SYSOUT=A,SPACE=(TRK,10,RLSE),DCB=BLKSIZE=2178
IEF236I ALLOC. FOR AGB005Z1 LKED
IEF237I 1C0   ALLOCATED TO SYSLIB
IEF237I 230   ALLOCATED TO SYSLIN
IEF237I 230   ALLOCATED TO SYSLMOD
IEF237I 230   ALLOCATED TO SYSUT1
IEF237I 630   ALLOCATED TO SYSPRINT
```

```
F128-LEVEL LINKAGE EDITOR OPTIONS SPECIFIED MAP
        DEFAULT OPTION(S) USED -  SIZE=(151552,40960)
```

MODULE MAP

CONTROL SECTION			ENTRY							
NAME	ORIGIN	LENGTH	NAME	LOCATION	NAME	LOCATION	NAME	LOCATION	NAME	LOCATION
MAIN	00	A18								
BLOCK8	A18	1340								
BMOMTS	1D58	A8C								
FRV	27E8	2E0								
GAMMA	2AC8	40C								
JKNIFE	2ED8	798								
LOAD	3670	2C0								
LOGNL	3930	1540								
LP	4E70	AF0								
NEGHGR	5960	740								
PEARSN	60A0	2A98								
POLRT	8B38	92E								
RDFMT	9468	28A								
SETUP	96F8	56C								
MOMVAL	9C68	470								
SUBTST	A0D8	148B4								
TTABLE	1E990	528								
VCOV	1EEB8	5DE								
IHCSLOG *	1F498	1B6	ALOG10	1F498	ALOG	1F4B0				
IHCSATN2*	1F650	1CB	ATAN2	1F650	ATAN	1F664				
IHCSSCN *	1F820	1D9	COS	1F820	SIN	1F838				
IHCFEXIT*	1FA00	1C	EXIT	1FA00						
IHCSEXP *	1FA20	192	EXP	1FA20						
IHCFRXPI*	1FBB8	141	FRXPI#	1FBB8						
IHCFRXPR*	1FD00	183	FRXPR#	1FD00						
IHCECOMH*	1FE88	F61	IBCOM#	1FE88	FDIOCS#	1FF44	INTSWTCH	20DCE		
IHCCOMH2*	20DF0	65D	SEQDASD	21168						
IHCSSQRT*	21450	145	SQRT	21450						
IHCFCVTH*	21598	119D	ADCON#	21598	FCVAOUTP	21642	FCVLOUTP	216D2	FCVZOUTP	21822
			FCVIOUTP	21BD0	FCVEOUTP	220D2	FCVCOUTP	222EC	INT6SWCH	225D3
IHCEFNTH*	22738	542	ARITH#	22738	ADJSWTCH	22AD4				
IHCLSQRT*	22C80	15B								

```
   NAME      ORIGIN  LENGTH            NAME    LOCATION      NAME    LOCATION      NAME   LOCATION      NAME   LOCATION
                                     DSQRT       22C80
  IHCEFIOS*  22DE0     EBC
                                     FIOCS#      22DE0      FIOCSBEP   22DE6
  IHCFIOS2*  23CA0     52E
  IHCERRM *  241D0     5D4
                                     ERRMON      241D0      IHCERRE    241E8
  IHCUOPT *  247A8     340
  IHCETRCH*  24AE8     28E
                                     IHCTRCH     24AE8      ERRTRA     24AF0
  IHCUATBL*  24D78     638
  BLOCK1     253B0    118C
  BLOCK5     26540    9DE0

 ENTRY ADDRESS       00
 TOTAL LENGTH     30320

****GOFORTG   DOES NOT EXIST BUT HAS BEEN ADDED TO DATA SET
```

```
IEF285I   SYS1.FORTLIB                                    PASSED
IEF285I   VCL SER NCS= DRUM01.
IEF285I   SYS73101.T070543.RV000.AGB005Z1.LOADSET         DELETED
IEF285I   VOL SER NOS= SYSDA1.
IEF285I   SYS73101.T070543.RV000.AGB005Z1.GCSET           PASSED
IEF285I   VOL SER NCS= SYSDA1.
IEF285I   SYS73101.T070543.RV000.AGB005Z1.R0000003        DELETED
IEF285I   VOL SER NOS= SYSDA1.
IEF285I   SYS73101.T070543.SV000.AGB005Z1.R0000004        SYSOUT
IEF285I   VOL SER NOS= SYSDA4.
CCN011I STEF RETURN CODE =    0

CCN301I   SYSPRINT PRINTED     2 TOTAL PAGES, REQUIRED     2 TRACKS.

  REGION ALLOCATION        CORE USED          MAIN STORAGE      EXTENDED STORAGE
         300K                150K                 300K                  0K

   STEP CPU TIME          JOB CPU TIME       STEP I/O COUNT      JOB I/O COUNT
        0.79S                16.68S               401                 751

XXGO      EXEC   PGM=*.LKED.SYSLMOD,COND=((5,LT,FORT),(5,LT,LKED)),        00000170
XX  REGION=&RG,TIME=&TG                                                     00000180
IEF653I SUBSTITUTION JCL - REGION=56K,TIME=1439
XXFT05F001 DD DDNAME=SYSIN                                                  00000190
XXSTEPLIB  DD DISP=(SHR,PASS),DSN=$YS&LEVEL..FORTRAN                        00000200
IEF653I SUBSTITUTION JCL - DISP=(SHR,PASS),DSN=$YS1.FORTRAN
XXFT06F001 DD SYSOUT=A,DCB=(BLKSIZE=3458,RECFM=FBA,LRECL=133),              00000210
XX  SPACE=(TRK,&PG,RLSE)                                                    00000220
IEF653I SUBSTITUTION JCL - SPACE=(TRK,50,RLSE)
//GO.FT02F001 DD DSN=&SCRATCH,UNIT=SYSDA,SPACE=(TRK,150,RLSE),
// DCB=(RECFM=FBA,BLKSIZE=7280,LRECL=80)
//GO.SYSIN DD *
//
IEF236I ALLCC. FOR AGB005Z1 GO
IEF237I 230   ALLOCATED TO PGM=*.DD
IEF237I 540   ALLOCATED TO FT05F001
IEF237I 342   ALLOCATED TO STEPLIB
IEF237I 630   ALLOCATED TO FT06F001
IEF237I 230   ALLOCATED TO FT02F001
```

```
CARD IMAGES LOADED ON SCRATCH FILE 2

01 1 1     1011111110
02 1 1     1111101011
03 1 1     1110100000
04 1 1     1110000000
05 1 1     0000000000
06 1 1     1101010001
07 2 1     1110101010
08 2 1     1100100000
09 2 1     0000011010
10 2 1     0100000000
11 2 1     1111101010
12 2 1     1001001000
13 3 1     0100100000
14 3 1     0111100010
15 3 1     0000000000
16 3 1     0000000000
17 3 1     1101001110
18 2 1     0000000000
01 1 2     1111110011
02 1 2     1011100000
03 1 2     1100101000
04 1 2     1111110000
05 1 2     1011110010
06 1 2     1110110000
07 2 2     1010100000
08 2 2     1111110010
09 2 2     1101111010
10 2 2     1111110000
11 3 2     0001110000
12 3 2     1111111100
13 3 2     1101000000
14 3 2     0011110000
01 1 3     1101111101
02 1 3     0101001000
03 1 3     1001011101
04 1 3     0010010000
05 1 3     1111111111
06 1 3     1101110100
07 2 3     0100000100
08 2 3     0111111110
09 2 3     1000000000
10 2 3     1001000000
11 3 3     1111111100
12 3 3     1111000100
13 3 3     0000000000
```

```
01 1 4    1101001100
02 1 4    0010001000
03 1 4    1111011011
04 1 4    1111111111
05 1 4    1111011111
06 2 4    0111010000
07 2 4    1111011111
08 2 4    0000000000
09 2 4    0110001100
10 3 4    0011001000
11 3 4    1111001110
12 3 4    0011001000
13 3 4    0000000000
01 1 5    1001001010
02 1 5    0000000000
03 1 5    1000000000
04 1 5    0000000000
05 2 5    0101111010
06 2 5    1110010100
07 2 5    0000000000
08 2 5    0000000000
09 3 5    1011111100
10 3 5    0000000000
11 3 5    0000000000
12 3 5    0000000000
END
```

```
FIRST YEAR WORD SPELLING PROJECT     SHOEMAKER/OKADA

SPECIFICATICNS FOR PROBLEM SET

NUMBER OF ITEMS            50
MAX ITEM SCORE              1
NUMBER OF SUBTESTS          5
ITEM SCORING                0
NORMATIVE DISTRIBUTION      1
STR2 OPTION USED

SUBTEST  EXAMINEES  ITEMS  FORMAT

   1        18       10    (10X,10F1.0)
                           ITEM DIFFICULTY INDICES
                             0.556  0.611  0.389  0.389  0.444  0.167  0.389  0.111  0.389  0.111
   2        14       10    (10X,10F1.0)
                           ITEM DIFFICULTY INDICES
                             0.857  0.643  0.714  0.786  0.929  0.714  0.214  0.071  0.286  0.071
   3        13       10    (10X,10F1.0)
                           ITEM DIFFICULTY INDICES
                             0.615  0.615  0.385  0.692  0.385  0.538  0.462  0.615  0.154  0.231
   4        13       10    (10X,10F1.0)
                           ITEM DIFFICULTY INDICES
                             0.462  0.615  0.769  0.692  0.077  0.385  0.769  0.462  0.385  0.308
   5        12       10    (10X,10F1.0)
                           ITEM DIFFICULTY INDICES
                             0.333  0.167  0.167  0.250  0.167  0.250  0.250  0.167  0.167  0.0
```

	ESTIMATES OF MOMENTS				COMPONENTS OF VARIANCE			
SUBTEST	MEAN	VARIANCE	U3	U4	VAR (I)	VAR (E)	VAR (IE)	THETA'
1	0.17778E 02	0.15651E 03	-0.25546E 02	0.38881E 05	0.21786E-01	0.60203E-01	0.15352E 00	0.39867E 00
2	0.26429E 02	0.60055E 02	0.21991E 03	-0.24597E 04	0.10000E 00	0.18681E-01	0.14286E 00	0.13478E 00
3	0.23462E 02	0.23308E 03	0.60330E 03	0.66050E 05	0.20655E-01	0.89031E-01	0.14943E 00	0.60893E 00
4	0.24615E 02	0.27109E 03	0.36145E 03	0.11413E 06	0.39744E-01	0.10385E 00	0.11923E 00	0.88929E 00
5	0.95833E 01	0.16609E 03	0.15471E 04	0.21412E 05	-0.16835E-03	0.65825E-01	0.95539E-01	0.70525E 00

SE(MEAN)-HLN = 2.10472
SE(VAR)-HLN = 24.46585

	MEAN	VARIANCE	U3	U4	VAR (I)	VAR (E)	VAR (IE)	THETA'
MEAN	0.20417E 02	0.17355E 03	0.46546E 03	0.46351E 05	0.36755E-01	0.66007E-01	0.13493E 00	0.52356E 00
SE	0.27593E 01	0.34912E 02	0.26099E 03	0.18796E 05	0.16713E-01	0.14039E-01	0.10458E-01	0.12696E 00
90.0 PERCENT CCNFIDENCE INTERVAL								
FROM	0.14533E 02	0.99108E 02	-0.91046E 02	0.62725E 04	0.11185E-02	0.36071E-01	0.11263E 00	0.25286E 00
TO	0.26300E 02	0.24799E 03	0.10220E 04	0.86429E 05	0.72392E-01	0.95943E-01	0.15723E 00	0.79426E 00
95.0 PERCENT CONFIDENCE INTERVAL								
FROM	0.12755E 02	0.76614E 02	-0.25920E 03	-0.58378E 04	-0.96496E-02	0.27026E-01	0.10589E 00	0.17106E 00
TO	0.28078E 02	0.27049E 03	0.11901E 04	0.98540E 05	0.83160E-01	0.10499E 00	0.16397E 00	0.87606E 00
97.5 PERCENT CONFIDENCE INTERVAL								
FROM	0.10776E 02	0.51571E 02	-0.44642E 03	-0.19321E 05	-0.21638E-01	0.16955E-01	0.98390E-01	0.79996E-01
TO	0.30057E 02	0.29553E 03	0.13774E 04	0.11202E 06	0.95148E-01	0.11506E 00	0.17147E 00	0.96713E 00
99.0 PERCENT CONFIDENCE INTERVAL								
FROM	0.77363E 01	0.13110E 02	-0.73394E 03	-0.40027E 05	-0.40050E-01	0.14884E-02	0.86869E-01	-0.59863E-01
TO	0.33097E 02	0.33399E 03	0.16649E 04	0.13273E 06	0.11356E 00	0.13053E 00	0.18299E 00	0.11070E 01

ESTIMATED RELIABILITY OF UNIVERSE SCORE IS 0.96321

MINIMUM AND MAXIMUM POSSIBLE VALUES FOR ESTIMATED MOMENTS

	ABOUT ORIGIN			ABOUT MEAN			
MOMENT	ESTIMATE	MINIMUM	MAXIMUM	ESTIMATE	MINIMUM	MAXIMUM	APPRAISAL
1	0.2041664E 02			0.0			
2	0.5903892E 03	0.4170820E 03	0.1020832E 04	0.1735501E 03	0.2429199E 00	0.6039927E 03	CK
3	0.1960584E 05	0.1707387E 05	0.2324908E 05	0.4654646E 03	-0.2066512E 04	0.4108699E 04	CK
4	0.6921724E 06	0.6772897E 06	0.7519546E 06	0.4635092E 05	0.3146725E 05	0.1061321E 06	OK

APPROXIMATION OF NORMATIVE DISTRIBUTION WITH NEGATIVE HYPERGEOMETRIC DISTRIBUTION

ESTIMATED PARAMETERS

MEAN	20.41664
VARIANCE	173.55009
A21	0.94938

INDEX	RELATIVE FREQ	CUM REL FREQ
0	0.02305	0.02305
1	0.02481	0.04786
2	0.02560	0.07346
3	0.02605	0.09951
4	0.02630	0.12581
5	0.02643	0.15224
6	0.02648	0.17872
7	0.02647	0.20519
8	0.02641	0.23160
9	0.02631	0.25790
10	0.02617	0.28407
11	0.02601	0.31008
12	0.02582	0.33590
13	0.02560	0.36150
14	0.02537	0.38687
15	0.02512	0.41198
16	0.02484	0.43683
17	0.02456	0.46139
18	0.02425	0.48564
19	0.02394	0.50958
20	0.02360	0.53318
21	0.02325	0.55643
22	0.02289	0.57932
23	0.02251	0.60184
24	0.02212	0.62396
25	0.02172	0.64568
26	0.02130	0.66699
27	0.02087	0.68786
28	0.02043	0.70828
29	0.01996	0.72825

```
       30            0.01949              0.74773
       31            0.01900              0.76673
       32            0.01849              0.78522
       33            0.01796              0.80317
       34            0.01741              0.82059
       35            0.01685              0.83744
       36            0.01627              0.85370
       37            0.01566              0.86936
       38            0.01503              0.88439
       39            0.01437              0.89876
       40            0.01368              0.91245
       41            0.01297              0.92541
       42            0.01221              0.93762
       43            0.01141              0.94903
       44            0.01056              0.95960
       45            0.00966              0.96925
       46            0.00867              0.97792
       47            0.00759              0.98552
       48            0.00638              0.99190
       49            0.00496              0.99686
       50            0.00315              1.00001
ALL INPUT PROCESSED


IEF285I   SYS73101.T070543.RV000.AGB005Z1.GOSET          PASSED
IEF285I   VCL SER NOS= SYSDA1.
IEF285I   SYS73101.T070543.RV000.AGB005Z1.S0000006       SYSIN
IEF285I   VOL SER NOS= SYSDA3.
IEF285I   SYS73101.T070543.RV000.AGB005Z1.S0000006       DELETED
IEF285I   VOL SER NOS= SYSDA3.
IEF285I   SYS1.FORTRAN                                   PASSED
IEF285I   VCL SER NOS= SYSTM3.
IEF285I   SYS73101.T070543.SV000.AGB005Z1.R0000005       SYSOUT
IEF285I   VOL SER NOS= SYSDA4.
IEF285I   SYS73101.T070543.RV000.AGB005Z1.SCRATCH        DELETED
IEF285I   VCL SER NCS= SYSDA1.
CCN011I STEF RETURN CODE =    0

CCN301I   FT06F001 PRINTED     6 TOTAL PAGES, REQUIRED     5 TRACKS.

  REGION ALLOCATION       CORE USED        MAIN STORAGE      EXTENDED STORAGE
        300K                230K               300K                 0K

   STEP CPU TIME        JOB CPU TIME      STEP I/O COUNT      JOB I/O COUNT
        1.08S              17.76S              143                 894

IEF285I   SYS1.FORTRAN                                   KEPT
IEF285I   VOL SER NOS= SYSTM3.
IEF285I   SYS1.FORTLIB                                   KEPT
IEF285I   VOL SER NOS= DRUM01.
IEF285I   SYS73101.T070543.RV000.AGB005Z1.GOSET          DELETED
IEF285I   VOL SER NOS= SYSDA1.


********** END OF JOB **********
```

Appendix B

Program for Simulating Multiple Matrix Sampling

1.0 - PROGRAM IDENTIFICATION

Program for simulating multiple matrix sampling.

2.0 - OBJECTIVE

This program simulates the testing of a random sample of examinees over a subset of items selected randomly from the item universe. The results obtained from replicating a particular sampling plan using a data base specified by the user may be used (a) to approximate the standard error of estimate associated with any estimate of a parameter, (b) to determine the feasibility of procedures devised for estimating parameters, and (c) to familiarize researchers with the types of results to be expected from multiple matrix sampling.

3.0 - PROGRAM DESCRIPTION

The program is designed to simulate multiple matrix sampling and to estimate concomitantly standard errors of estimate associated with a variety of parameters describing the data base. The general algorithm used within the program involves five main steps: (a) the generation of universe scores and item scores having characteristics defined by the user, (b) sampling randomly items and examinees from this data base using a sampling plan specified by the user, (c) computing estimates of parameters from subtest results, (d) repeating (replicating) steps (b) and (c) as many times as specified by the user, and (e) computing the standard deviation over replications of the pooled (over subtests) estimates made of each parameter.

To initialize the program, the user must specify five parameters of the data base: (a) the number of items in the universe, (b) the mean universe score, (c) the variance of the item difficulty indices, (d) the degree of skewness in the distribution of universe scores, and (e) either the variance or the reliability of the universe scores. Although the general procedure used by the program is postmortem item-examinee sampling, the implementation of the procedure differs from that encountered typically. The standard procedure for postmortem sampling is to begin initially with an entire data base stored in memory; however, in this program the universe score and item scores are generated individually for each examinee and the only scores in memory are those generated for the last examinee selected within the program. The procedure for generating the data base is one such that the item scores and universe scores for all examinees generated have parameters as specified by the user.

The universe score for each examinee is sampled randomly from either a normal or lognormal distribution. In rounding a continuous score to an integer score (number correct score), the universe score between each integer test score interval dividing equally the area within the interval is determined and continuous scores above this point are rounded to the next higher integer with all others rounded to the next lower integer. All items are "scored" dichotomously (1 = pass, 0 = fail) and the generation of scores for each item per examinee is done so as to make the item difficulty indices computed on the data generated for all previous examinees more closely approximate those of the data base specified by the user.

The parameters of the data base estimated at each replication of the sampling plan are (a) the mean universe score, (b) the second, third, and fourth central moments of universe scores, (c) components of variance for items, examinees, and the item by examinee interaction, (d) the average reliability of a single item, (e) the 90, 95, 97.5, and 99 percent confidence interval for parameters (a) through (d), and (f) the relative normative distribution. In computing the confidence intervals for each parameter, the jackknife procedure is used in conjunction with the *t*-distribution having degrees of freedom equal to the number of subtests minus one. Because the jackknife procedure approximates at each replication of the sampling plan the standard error of estimate which is computed normally by the standard deviation of the pooled estimates over replications, the statistics of the jackknife estimation procedure are included in the summary of results obtained over replications. Additionally, the standard error of the pooled estimate of the mean universe score and variance of universe scores is computed using the Hooke-Lord-Novick equations.

4.0 - SUBROUTINES AND FUNCTIONS

4.1 - Coded Subprograms

ALLOC - Allocation of items in universe to subtests according to item-sampling plan specified by user.

SUBROUTINE ALLOC (NT,IPT,ISAMP,KPOP)

NT - Number of subtests
IPT - Number of items per subtest
ISAMP - Code number of item-sampling plan
KPOP - Number of items in universe

BMOMTS - Determination of minimum and maximum values for each estimated moment of universe scores.

SUBROUTINE BMOMTS (V1,U2,U3,U4,KPOP)

V1 - Estimated mean universe score
U2 - Estimated variance of universe scores
U3 - Estimated third central moment
U4 - Estimated fourth central moment
KPOP - Number of items in universe

DATA - Generation of item scores and universe score for randomly selected examinee.

SUBROUTINE DATA (X)

X - Vector containing item scores on each item in universe for examinee selected

ESTORS - Calculation of estimates of parameters from results of one individual subtest.

SUBROUTINE ESTORS (XX,NSPT,IPT,Z,VU1,VU2)

XX - Examinee by item matrix containing item scores
NSPT - Number of examinees (rows)
IPT - Number of items (columns)
Z - Vector containing estimates of parameters obtained from individual subtest

VU1 - Estimate of squared standard error of mean
VU2 - Estimate of squared standard error of variance

FRV – Evaluation of $\cos(x)^r e^{-vx} dx$ from -1.570796 to 1.570796 using Simpson's rule.

FUNCTION FRV (R,VV)

R - 6.*(B2-B1-1.)/(2.*B2-3.*B1-6.)
VV - SQRT((R*R*(R-2.)**2*B1)/(16.*(R-1.)-B1*(R-2.)**2))

GAMMA - Computation of gamma function using the recursion relation and polynomial approximation.

FUNCTION GAMMA (XX)

XX - Any real argument

LP - Linear programming subroutine. (Subroutine written originally by Dr. James E. Martin, Virginia Polytechnic Institute, and modified subsequently by Dr. Richard K. Hill, Jr., Virginia Polytechnic Institute, and David M. Shoemaker.)

SUBROUTINE LP (NERR)

NERR - Error return code

MOMVAL - Conversion of frequencies in X-matrix from alphanumeric to numeric and calculation of moment value. (Subroutine written originally by Dr. Richard K. Hill, Jr., Virginia Polytechnic Institute, and modified subsequently by David M. Shoemaker.)

SUBROUTINE MOMVAL (VAL,MOM,VALC)

VAL - Value of moment about origin
MOM - Degree of moment
VALC - Value of moment about mean

NEGHGR - Graduation of normative distribution by negative hypergeometric distribution.

SUBROUTINE NEGHGR (KPOP,XBAR,XVAR)

KPOP - Number of items in universe
XBAR - Mean universe score
XVAR - Variance of universe scores

PEARSN - Graduation of normative distribution with one of the family of Pearson curves.

SUBROUTINE PEARSN (U1,U2,U3,U4,L1,L2,STEP)

U1 - Mean universe score
U2 - Second central moment
U3 - Third central moment
U4 - Fourth central moment
L1 - Lower limit of index
L2 - Upper limit of index
STEP - Desired step size in going from L1 to L2

POLRT - Computation of the real and complex roots of a real polynomial.

SUBROUTINE POLRT (XCOF,M,ROOTR,ROOTI,IER)

XCOF - Vector of $M + 1$ coefficients of the polynomial ordered from smallest to largest power
M - Order of polynomial
ROOTR - Resultant vector of length M containing real roots
ROOTI - Resultant vector of length M containing the corresponding imaginary roots
IER - Error return code

POOL - Calculation of pooled estimates of parameters from results obtained from individual subtests.

SUBROUTINE POOL (NSPT,IPT,NT,Z,SE,S1,S2)

NSPT - Number of examinees per subtest
IPT - Number of items per subtest
NT - Number of subtests
Z - Vector containing pooled estimates of parameters

SE - Vector containing estimates of the standard errors obtained through the jackknife procedure
S1 - Standard error of pooled mean obtained from Hooke-Lord-Novick equation
S2 - Standard error of pooled variance obtained from Hooke-Lord-Novick equations

PVALUE - Generation of item difficulty indices (proportion of examinees answering correctly each item) having specified mean and variance.

SUBROUTINE PVALUE (PVAR)

PVAR - Variance of item difficulty indices

RANDND - Generation of random standard normal ND(0,1) deviate.

SUBROUTINE RANDND (X)

X - Random standard normal deviate generated

ROUND - Determination of the appropriate rounding value for each test score interval. The area between two adjacent test scores is computed by means of the trapezoid formula. The rounding value is that continuous test score such that one-half of the area within the score interval is above that point.

SUBROUTINE ROUND

(All arguments are in COMMON.)

RUNIF - Generation of uniform random numbers (.00 to 1.00) by the power residue technique. (Subroutine designed specifically for the IBM S360.)

SUBROUTINE RUNIF (R)

R - Uniform random number generated

SETUP - Preparation of *X*-matrix for use by LP subroutine. (Subroutine written originally by Dr. Richard K. Hill, Jr., Virginia Polytechnic Institute.)

SUBROUTINE SETUP (NI,NJ,NK,XMULT)

NI - Degree of moment
NJ - Number of items in universe plus seven
NK - Degree of moment
XMULT - Switch indicating whether minimum or maximum boundary is to be computed

TMOMTS – Calculation of first moment and second through fourth central moments for truncated normative distribution. The computed moments are those calculated on 5000 values sampled randomly from the normative distribution specified by the user.

SUBROUTINE TMOMTS (TU1,TU2,TU3,TU4)

TU1 - Estimate of mean universe score for truncated normative distribution
TU2 - Estimate of second central moment
TU3 - Estimate of third central moment
TU4 - Estimate of fourth central moment

TTABLE - Calculation of Student's *t* corresponding to prescribed alpha level (two-tailed probability).

SUBROUTINE TTABLE (T,N,AREA)

T - Student's *t*
N - Degrees of freedom associated with *t*
AREA - Relative area associated with confidence interval

VCOV - Calculation of variance of item-test covariances for subtest

SUBROUTINE VCOV (XX,NSPT,IPT,VCOVIT)

XX - Examinee by item matrix containing item scores
NSPT - Number of examinees
IPT - Number of items
VCOVIT - Variance of item-test covariances

4.2 - Library Subprograms

ALOG	AMIN1
ALOG10	AMAX1
EXP	ATAN
EXIT	COS
SQRT	

5.0 - DATA SPECIFICATIONS

5.1 - Input Formats

Input formats are described in detail at the beginning of the program listing. An example of an input data structure is given in Attachment 1.

5.2 - Output Formats

Output resulting from the input data structure listed in Attachment 1 is given in Attachment 2.

6.0 - PROGRAM CONSTRAINTS AND LIMITATIONS

6.1 - Programming Language
Fortran IV

6.2 - Storage Requirements
Total core for execution: 300K

6.3 - Hardware Configuration
IBM 360 Model 91, Punched Cards

6.4- Program Parameters

6.4.1 - The maximum permitted number of items in the universe is 150.
6.4.2 - All items are scored dichotomously.
6.4.3 - The product of the number of subtests and number of items per subtest cannot exceed 1500.

6.5 - Error Messages
The following error messages may be generated by the program:

1. Error in *t*-table arg: df = _ (TTABLE)

2. Error in *t*-table arg: *p* = _	(TTABLE)
3. Invalid limits and/or delta arguments	(PEARSN)
4. Ordinate undefined at this point	(PEARSN)
5. M .LE. 1. for Type XI	(PEARSN)
6. M out of range for Type VIII	(PEARSN)
7. M negative for Type IX	(PEARSN)
8. Gamma arg within .0000001 of being a negative integer	(GAMMA)
9. Gamma arg GT 57, overflow, ans set to 1.E75	(GAMMA)
10. Error on parameter card or need more information on parameters	(MAIN)
11. TK not integer multiple of K. Item-sampling switch reset to _	(ALLOC)
12. TK exceeds 1500, K reset to _	(ALLOC)
13. TK exceeds 1500, T reset to _	(ALLOC)
14. Problem in ROUND Subroutine. *X* =_	(ROUND)
15. No solution for_____moment	(LP)

7.0 - OPERATING INSTRUCTIONS

Submit program in batch mode in the following form:

```
col 1
// job card
//  PASSWORD
//  EXEC FORTGCLG,REGION=300K
// FORT.SYSIN DD *

main
subroutines

/*
// GO.SYSIN DD *

data structures

/*
//
```

8.0 PROGRAM FLOW CHART (MACRO)

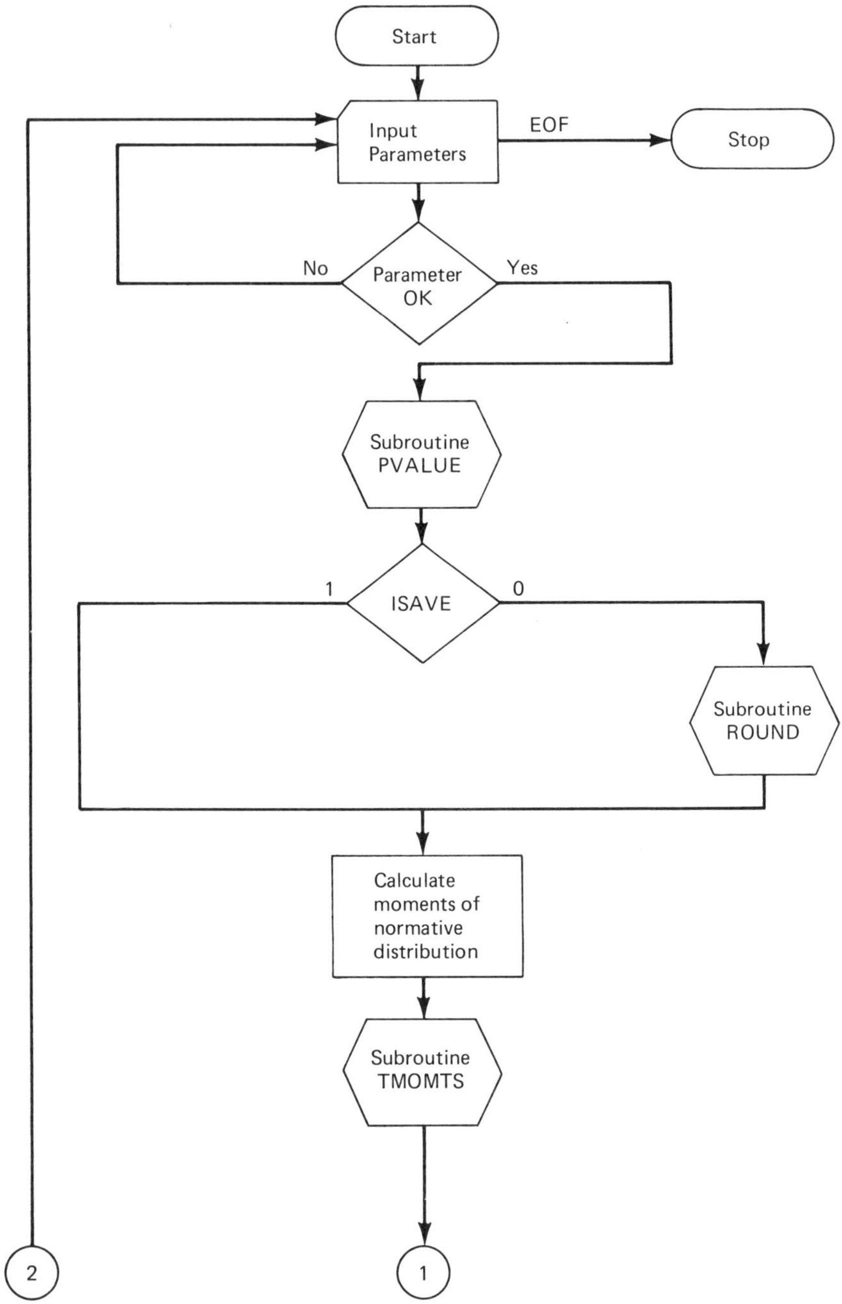

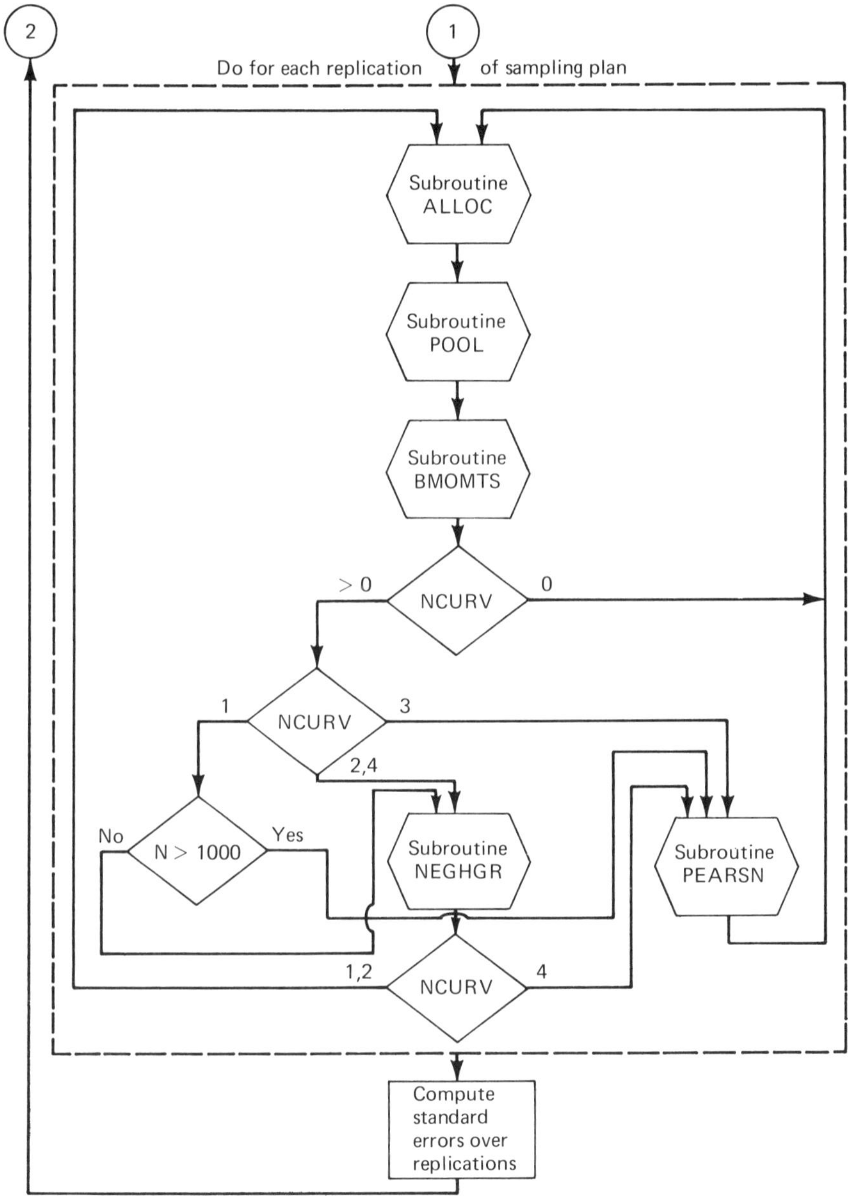
2
1
Do for each replication
of sampling plan
Subroutine ALLOC
Subroutine POOL
Subroutine BMOMTS
> 0
NCURV
0
1
NCURV
3
2,4
No
N > 1000
Yes
Subroutine NEGHGR
Subroutine PEARSN
1,2
NCURV
4
Compute standard errors over replications

```
FORTRAN IV G LEVEL  21            MAIN          DATE = 73097        11/11/21          PAGE 0001

C
C  DATE OF MOST RECENT REVISICN: 6 APRIL 73
C
C***********************************************************************
C
C  COMPUTER SIMULATION OF MULTIFLE MATRIX SAMFLING
C
C  LAVID M. SHOEMAKER
C  SOUTHWEST REGIONAL LABORATORY FOR EDUCATIONAL
C  RESEARCH AND DEVELOFMENT
C
C
C  PARAMETER CARD (THERE IS JUST ONE)
C
C     COLUMNS         (ALL INTEGEFS RIGHT-JUSTIFIED)
C     -------         -----------------------------------------------------
C
C      01-03          INTEGER NUMBER OF ITEMS IN UNIVERSE
C
C      04-09          DESIRED MEAN UNIVERSE SCORE IN FOPULATION
C                     (MUST BE SPECIFIED, WITH DECIMAL POINT PUNCHED ON
C                     CARD)
C
C      10-15          DESIRED VARIANCE OF UNIVERSE SCORES IN POPULATION
C                     (WITH DECIMAL POINT PUNCHED ON CARD)
C
C                     NOTE ... IF VARIANCE IS OMITTED, RELIABILITY MUST
C                     BE SPECIFIED.
C
C      16-21          DESIRED VARIANCE OF ITEM DIFFICULTY INDICES OVER
C                     POPULATION OF EXAMINEES.  THE ITEM DIFFICULTY
C                     INDEX FOR ITEM I IS THE PROPORTION OF EXAMINEES
C                     ANSWERING ITEM I CORRECTLY.
C                     (MUST BE SPECIFIED, WITH DECIMAL POINT PUNCHED ON
C                     CARD)
C
C                     WITH SKEWED DISTRIBUTIONS, VARIANCE OF ITEM
C                     DIFFICULTY INDICES IS ASSUMED TO BE EQUAL TO ZERO.
C
C
C      22-27          DESIRED RELIABILITY OF UNIVERSE SCORES (KR-20)
C                     (WITH DECIMAL POINT PUNCHED ON CARD)
C
C                     NOTE ... IF RELIABILITY IS OMITTED, VARIANCE MUST
C                     BE SPECIFIEL.
C
C      28-31          INTEGER NUMBEB OF SUBTESTS IN MULTIPLE MATRIX
C                     SAMPLING PLAN
C
C      32-35          INTEGER NUMBEB OF ITEMS PER SUBTEST
C                     (CONSTANT ACROSS SUBTESTS)
C
C      36-39          INTEGER NUMBEB OF EXAMINEES PER SUBTEST
C                     (CONSTANT ACROSS SUBTESTS)
```

```
FORTRAN IV G LEVEL  21          MAIN          DATE = 73097          11/11/21          PAGE C002
C
C      40-43     INTEGER NUMBER OF INDEPENDENT REPLICATIONS OF
C                MULTIPLE MATRIX SAMPLING PLAN
C
C        44      SAMPLING PLAN FOR ITEMS
C
C                     0 = SAMPLING WITH REPLACEMENT ACROSS SUBTESTS
C
C                     1 = SAMPLING WITHOUT REPLACEMENT ACROSS SUBTESTS
C                         (USED WHEN TK IS LESS THAN K)
C
C
C                     2 = SAMPLING WITHOUT REPLACEMENT BUT SUBJECT
C                         TO RESTRICTION THAT ITEMS OCCUR WITH EQUAL
C                         FREQUENCY AMONG SUBTESTS
C                         (USED WHEN TK IS GREATER THAN OR EQUAL TO K)
C
C        45      INTERMEDIATE PRINTOUT OPTION
C
C                     0 = NO INTERMEDIATE PRINTOUT
C                     1 = INTERMEDIATE PRINTOUT WANTED
C
C        46      CALCULATE NORMATIVE DISTRIBUTION
C
C                     0 = NO NORM DIST WANTED
C                     1 = DIST WANTED (DIST SELECTED BY PROGRAM)
C                     2 = DIST WANTED (NEGATIVE HYPERGEOMETRIC)
C                     3 = DIST WANTED (PEARSON CURVE)
C                     4 = DIST WANTED (BOTH NEG HYPER AND PEARSON  CURVES)
C
C
C        47      DEGREE OF SKEWNESS IN NORMATIVE DISTRIBUTION
C
C                     1 = NORMALLY DISTRIBUTED
C                     2 = POSITIVELY SKEWED
C                     3 = NEGATIVELY SKEWED
C
C
C      50-53     SEED FOR UNIFORM RANDOM NUMBER GENERATOR
C                (ANY POSITIVE INTEGER)
C
C        55      GENERATE ITEM DIFFICULTY INDICES
C
C                     0 = GENERATE NEW ITEM DIFFICULTY INDICES
C                     1 = USE ITEM DIFFICULTY INDICES GENERATED BY
C                         PREVIOUS DATA CARD
C
C        57      GENERATE FREQUENCY DISTRIBUTIONS FOR POOLED
C                PARAMETER ESTIMATES
C
C                     0 = NO FREQ. DIST. WANTED
C                     1 = TABULATE FREQUENCY DISTRIBUTIONS
C
C  RESTRICTIONS
```

```
FORTRAN IV G LEVEL  21                MAIN              DATE = 73097        11/11/21              PAGE 0003

                 C
                 C      MAXIMUM NUMBER OF ITEMS IN UNIVERSE IS 150  (EASILY MODIFIED)
                 C
                 C      PRODUCT OF NUMBER OF SUBTESTS AND NUMBER OF ITEMS PER SUBTEST
                 C      CANNOT EXCEED 1500
                 C
                 C      ITEMS SCORED DICHOTOMOUSLY
                 C
                 C
                 C   PROGRAM WILL PROCESS REPEATED PARAMETER CARDS (NUMBER LIMITED ONLY
                 C   BY AMCUNT OF COMPUTER TIME ALLOCATED)
                 C
                 C***********************************************************************
                 C
0001                    INTEGER BASE,FQ
0002                    REAL MCMT1,MOMT3,MCMT4,N,M,MPOP
0003                    COMMON N(150),M(150),LT(1500)
0004                    CCMMCN /BLOCK1/ YEAR,YSD,MPOP,SPOP,KPOP,NDIST,INTPRT
0005                    COMMON /BLOCK2/ RND(150)
0006                    COMMON /BLOCK3/ P(150),Q(150),NSUB
0007                    COMMON /BLOCK5/ BASE,COSU(290)
0008                    COMMCN /BLOCK7/ FQ(52,13),CFQ(5,13),NFREQ
0009                    DIMENSION SX(8),SXX(8),Z(13),SE(13),SEBAR(8),SEVAR(8),CC(4,7)
0010                    EQUIVALENCE (Z(1),XBAR),(Z(2),XVAR),(Z(3),U3),(Z(4),U4),
                       1      (Z(5),PVR),(Z(6),THETA)
                 C
                 C
0011              2000 READ (5,1,END=5000)  KPOP,MPOP,VPOP,PVAR,A20,NT,IPT,NSPT,NREPS,
                       1                    ISAMP,INTPRT,NCURV,NDIST,BASE,ISAVE,NFREQ
0012                    BASE=(BASE/2)*2+1
0013                    NBS=BASE
0014                    IF ( NDIST .GT. 1 ) PVAR=0.
                 C
                 C
0015                    NSUB=0
                 C
                 C   CHECK CN PARAMETERS
                 C
0016                    IF ( ISAMP .NE. 1 ) GO TO 30
0017                    IF ( NT*IPT .GT. KPOP ) GC TO 55
0018               30   IF ( A20 .LT. 0. .OR. A20 .GT. 1. ) GO TO 55
0019                    IF ( NDIST .GT. 4 .OR. NDIST .LT. 1 ) GO TO 55
0020                    IF ( A20 .EQ. 0. .AND. VPOP .EQ. 0. ) GC TO 55
0021                    IF ( MPOP .GE. FLCAT(KPOP) ) GO TO 55
0022                    IF ( PVAR .LT. .2 ) GO TO 70
0023               55   WRITE (6,3)
0024                    GC TC 2000
                 C
                 C   COMPUTE NECESSARY PARAMETERS
                 C
0025               70   TEMP=MPOP*(KPOP-MPOP)-KPCP*KPOP*PVAR
0026                    IF ( VPOP .EQ. 0. ) VPOP=TEMP/(KPOP-(KPOP-1.)*A20)
0027                    SPOP=SQRT(VPOP)
0028                    IF ( A20 .EQ. 0. ) A20=(KPOP*VPOP-TEMP)/((KPOP-1.)*VPOP)
```

```
FORTRAN IV G LEVEL  21              MAIN              DATE = 73097          11/11/21                 PAGE 0004

0029            THETA=A20/(KPOP-A20*KPOP)
0030            WRITE (6,2)  BASE,MPOP,NT,ISAMP,VPOP,IPT,NCURV,KPOP,NSPT,NDIST,
              1      A20,NREPS,INTPRT,PVAR,THETA
          C
          C  GENERATE ITEM DIFFICULTY INDICES (PROPORTION OF EXAMINEES ANSWERING
          C  ITEM CORRECTLY)
          C
0031            IF ( ISAVE .EQ. 0 ) CALL PVALUE(PVAR)
0032            IF ( INTPRT .EQ. 1 ) WRITE (6,4)  (P(I),I=1,KPOP)
          C
          C  COMPUTATION OF CONSTANTS FOR GENERATION OF LOGNORMAL DISTRIBUTION
          C
0033            YSD=0.
0034            IF ( NDIST .EQ. 1 ) GO TO 111
0035            IF ( NDIST .EQ. 3 ) MPOP=KPOP-MPOP
0036            YVAR=ALOG(VPOP/(MPOP*MPOP)+1.)
0037            YBAR=ALOG(MPOP)-YVAR/2.
0038            YSD=SQRT(YVAR)
          C
          C  COMPUTE ROUNDING VALUES FOR EACH TEST SCORE INTERVAL
          C
0039       111  IF ( ISAVE .EQ. 0 ) CALL ROUND
          C
          C
          C  CALCULATION OF THIRD AND FOURTH CENTRAL MOMENTS FOR NORMATIVE
          C  DISTRIBUTIONS
          C
0040            MOMT1=MPOP
0041            GO TO (91,92,92), NDIST
          C
          C  MOMENTS FOR NORMAL DISTRIBUTION
          C
0042        91  MOMT3=0.
0043            MOMT4=3.*VPOP*VPOP
0044            GO TO 94
          C
          C  MOMENTS FOR POSITIVELY-SKEWED LOGNORMAL DISTRIBUTION
          C
0045        92  ETA2=EXP(YVAR)-1.
0046            MOMT3=(MPOP**3)*(ETA2**3+3.*ETA2**2)
0047            MOMT4=(MPOP**4)*(ETA2**6+6.*ETA2**5+15.*ETA2**4
              1      +16.*ETA2**3+3.*ETA2**2)
0048            IF ( NDIST .EQ. 3 ) GO TO 93
0049            GO TO 94
          C
          C  MOMENTS FOR NEGATIVELY-SKEWED LOGNORMAL DISTRIBUTION
          C
0050        93  MOMT1=KPOP-MPOP
0051            MOMT3=-MOMT3
          C
          C
0052        94  CONTINUE
          C
          C  CALCULATE MOMENTS OF TRUNCATED NORMATIVE DISTRIBUTION
```

```
FORTRAN IV G LEVEL  21              MAIN              DATE = 73097          11/11/21              PAGE C005

                  C
 0C53                    CALL TMCMTS (TU1,TU2,TU3,TU4)
                  C
                  C
 0054                    WRITE (6,14)  MOMT1,TU1,VPOP,TU2,MOMT3,TU3,MOMT4,TU4
                  C
                  C   INITIALIZE FQ AND CFQ MATRICES USED IN TABULATING FREQUENCY
                  C   DISTRIBUTIONS OF POOLED ESTIMATES OF PARAMETERS
                  C
 0055                    IF ( NFREQ .EQ. 0 ) GO TO 105
 0056                    DO 96 I=1,13
 0057                    DC 96 J=1,52
 0058              96    FQ(J,I)=0
 0059                    CFQ(1,1)=MOMT1-KPOP/8.
 0060                    CFQ(2,1)=MCMT1+KPOP/8.
 0C61                    IF  ( CFQ(1,1)  .LT. 0. )  CFQ(1,1)=0.
 0062                    IF  ( CFQ(2,1)  .GT. FLOAT(KPOP) )  CFQ(2,1)=KPOP
 0063                    CFQ(5,1)=50.
 0064                    CFQ(4,1)=CFQ(2,1)-CFQ(1,1)
 0065                    CALL BMOMTS (U1,U2,U3,U4,KPOP,CC)
 0066                    CC(2,6)=CC(2,6)/3.
 0C67                    DO S7 I=2,4
 0068                    T1=CC(I,5)
 0069                    T2=CC(I,6)
 0C70                    CFQ(1,I)=T1
 0C71                    CFQ(2,I)=T2
 0072                    CFQ(5,I)=50.
 0073                    IF  ( T1 .GE. 0. ) GO TO 99
 0C74                    CFQ(4,I)=ABS(T1)+T2
 0C75                    GO TO 97
 0C76              99    CFQ(4,I)=T2-T1
 0077              97    CONTINUE
 0078                    DO 101 I=5,8
 0C79                    CFQ(5,I)=50.
 0080                    CFQ(1,I+5)=-5.
 0C81                    CFQ(2,I+5)=5.
 0082                    CFQ(4,I+5)=10.
 0083             101    CFQ(5,I+5)=50.
 0C84                    CFQ(1,5)=PVAR-.02
 0085                    CFQ(2,5)=PVAR+.02
 0086                    CFQ(4,5)=.04
 0C87                    CFQ(1,6)=-.2
 0088                    CFQ(2,6)=.8
 0089                    CFQ(4,6)=1.
 0090                    CFQ(1,7)=1.-A20-.1
 0091                    CFQ(2,7)=1.-A20+.1
 0C92                    CFQ(4,7)=.2
 C093                    TEMP=(VPOP-KPOP*(1.-A20))/KPOP**2
 C094                    CFQ(1,8)=TEMP-.1
 0095                    CFQ(2,8)=TEMP+.1
 0096                    CFQ(4,8)=.2
 0097                    CFQ(1,9)=0.
 0098                    CFQ(2,9)=1.
 0099                    CFQ(4,9)=1.
```

```
FORTRAN IV G LEVEL  21                MAIN              DATE = 73097          11/11/21             PAGE 0006

 0100                 CFQ(5,9)=50.
 0101                 DO 102 I=1,13
 0102           102   CFQ(3,I)=CFQ(4,I)/CFQ(5,I)
              C
              C
 0103           105   CONTINUE
              C
              C  REPLICATION OF MULTIPLE MATRIX SAMPLING PARADIGM
              C
 0104                 DO 95 I=1,8
 0105                 SEBAR(I)=0.
 0106                 SEVAR(I)=0.
 0107                 SX(I)=0.
 0108            95   SXX(I)=0.
 0109                 SRTT=0.
 0110                 S2RTT=0.
 0111                 S1L=0.
 0112                 S2L=0.
 0113                 S1L2=0.
 0114                 S2L2=0.
              C
              C
 0115                 DO 7000 IJK=1,NREPS
              C
              C
 0116                 CALL ALLOC (NT,IPT,ISAMP,KPOP)
              C
              C
 0117                 IF ( INTPRT .EQ. 0 ) GO TO 113
 0118                 WRITE (6,5)
 0119                 J=0
 0120                 N1=NT*IPT
 0121                 DO 112 I=1,N1,IPT
 0122                 KK=I+IPT-1
 0123                 J=J+1
 0124           112   WRITE (6,6) J,(LT(K),K=I,KK)
              C
              C
 0125           113   CALL POOL (NSPT,IPT,NT,Z,SE,S1,S2)
              C
              C
 0126                 S1L=S1L+S1
 0127                 S2L=S2L+S2
 0128                 S1L2=S1L2+S1*S1
 0129                 S2L2=S2L2+S2*S2
 0130                 TEMP=Z(8)
 0131                 Z(8)=Z(6)
 0132                 Z(6)=TEMP
 0133                 TEMP=SE(8)
 0134                 SE(8)=SE(6)
 0135                 SE(6)=TEMP
 0136                 IF ( SE(1) .GT. 0. ) Z(10)=(Z(1)-U1)/SE(1)
 0137                 IF ( SE(2) .GT. 0. ) Z(11)=(Z(2)-U2)/SE(2)
 0138                 IF ( SE(3) .GT. 0. ) Z(12)=(Z(3)-U3)/SE(3)
```

```
FORTRAN IV G LEVEL  21                MAIN             DATE = 73097         11/11/21                PAGE 0007

0139               IF ( SE(4) .GT. 0. ) Z(13)=(Z(4)-U4)/SE(4)
             C
             C  CALCULATION OF RELIABILITY OF TOTAL TEST SCORE ON KPOP-ITEM TEST
             C
0140               RTT=KPOP*THETA/(1.+KPOP*THETA)
0141               SRTT=SRTT+RTT
0142               S2RTT=S2RTT+RTT*RTT
0143               Z(9)=RTT
             C
0144               IF ( NFREQ .EQ. 1 ) WRITE (6,25) (Z(J),J=9,13)
             C
             C
0145               IF ( INTPRT .EQ. 1 ) WRITE (6,7) IJK,XBAR,SE(1),XVAR,SE(2),U3,
                  1     SE(3),U4,SE(4),PVR,SE(5),THETA,SE(6),RTT
             C
0146               IF ( NFREQ .EQ. 1 ) CALL FREAKS(Z)
             C
0147               DO 125 I=1,6
0148               SEVAR(I)=SEVAR(I)+SE(I)*SE(I)
0149               SEBAR(I)=SEBAR(I)+SE(I)
0150               SX(I)=SX(I)+Z(I)
0151           125 SXX(I)=SXX(I)+Z(I)*Z(I)
             C
             C  TEST FEASIBILITY OF ESTIMATED MOMENTS THROUGH LINEAR PROGRAMMING
             C
0152               IF ( NCURV .EQ. 0 ) GO TO 7000
0153               CALL BMOMTS (XBAR,XVAR,U3,U4,KPOP,CC)
             C
             C  GRADUATE NORMATIVE DISTRIBUTICN WITH PARAMETERS ESTIMATED THROUGH
             C  MULTIPLE MATRIX SAMPLING
             C
0154               S=KECP
0155               GO TO (120,130,140,150), NCURV
0156           120 IF ( NT*NSPT .GT. 1000 ) GO TO 140
0157           130 CALL NEGHGR (KPOP,XBAR,XVAR)
0158               GO TO 7000
0159           140 CALL PEARSN (XBAR,XVAR,U3,U4,0.,S,1.)
0160               GO TO 7000
0161           150 CALL NEGHGR (KPOP,XBAR,XVAR)
0162               CALL PEARSN (XBAR,XVAR,U3,U4,0.,S,1.)
             C
             C
0163          7000 CCNTINUE
             C
             C  COMPUTE STANDARD ERRCRS OF ESTIMATE OVER REPLICATIONS
             C
0164               TEMP=(S1L2-S1L**2/NREPS)/NREPS
0165               S1L2=0.
0166               IF ( TEMP .GT. 0. ) S1L2=SQRT(TEMP)
0167               TEMP=(S2L2-S2L**2/NREPS)/NREPS
0168               S2L2=0.
0169               IF ( TEMP .GT. 0. ) S2L2=SQRT(TEMP)
0170               S1L=S1L/NREPS
0171               S2L=S2L/NREPS
```

```
FORTRAN IV G LEVEL  21              MAIN              DATE = 73097          11/11/21                    PAGE 0008

0172               IF ( NREPS .LE. 1 ) GO TO 270
0173               DO 200 I=1,6
0174               SEVAR(I)=SQRT((SEVAR(I)-SEBAR(I)**2/NREPS)/NREPS)
0175               SEBAR(I)=SEBAR(I)/NREPS
0176               SXX(I)=SQRT((SXX(I)-SX(I)*SX(I)/NREPS)/NREPS)
0177           200 SX(I)=SX(I)/NREPS
0178               K=0
0179               DO 250 I=1,21,4
0180               K=K+1
0181               N(I)=SX(K)
0182               N(I+1)=SXX(K)
0183               N(I+2)=SEBAR(K)
0184           250 N(I+3)=SEVAR(K)
0185               TEMP=(S2RTT-SRTT**2/NREPS)/NREPS
0186               S2RTT=0.
0187               IF ( TEMP .GT. 0. ) S2RTT=SQRT(TEMP)
0188               SRTT=SRTT/NREPS
             C
             C
0189               IF ( INTPRT .EQ. 0 ) GO TO 260
0190               WRITE (6,2)  NBS,MOMT1,NT,ISAMP,VPOP,IPT,NCURV,KPOP,NSPT,NDIST,
                  1            A20,NREPS,INTEFT,PVAR,THETA
0191               WRITE (6,14)  MOMT1,TU1,VPOP,TU2,MOMT3,TU3,MOMT4,TU4
0192           260 WRITE (6,11)  (N(I),I=1,24)
0193               WRITE (6,18)  SRTT,S2RTT
0194           270 IF (NT*IPT.LE.KPOP.AND.ISAMP.NE.0) WRITE(6,20) S1L,S1L2,S2L,S2L2
0195               IF ( INTPRT .EQ. 1 ) WRITE (6,19)  (Q(I),I=1,KPOP)
             C
0196               IF ( NFREQ .EQ. 1 ) CALL FRKOUT (NREPS)
             C
0197               GO TO 2000
             C
             C  EXIT GRACEFULLY
             C
0198          5000 WRITE (6,12)
0199               CALL EXIT
             C
             C  FORMAT STATEMENTS
             C
0200            1  FORMAT (I3,4F6.0,4I4,4I1,I6,2I2)
0201            2  FORMAT (12H1PROBLEM NO.I7//20H PARAMETERS INPUTTED,5X,
                  127HITEM-EXAMINEE SAMPLING PLAN,4X,8HSWITCHES//2X,4HMEAN,F14.3,
                  26X,15HNO. OF SUBTESTSI11,5X,18HITEM SAMPLING PLANI7/
                  32X,3HVARF15.3,6X,17HITEMS PER SUBTESTI9,5X,
                  418HRELATIVE FREQ DISTI7/2X,1HK,I13,10X,21HEXAMINEES PER SUBTESTI5,
                  55X,14HNORMATIVE DISTI11
                  6/2X,8HKR-20   F10.3,6X,19HNO. OF REPLICATIONSI7,5X,
                  721HINTERMEDIATE PRINTOUTI4/2X,6HVAR(I)F12.3/
                  82X,5HTHETAF13.3//)
0202            3  FORMAT (28H *** ERROR ON PARAMETER CARD///12X,2HOR//29H NEED MORE
                  1INFO ON PARAMETERS)
0203            4  FORMAT (    24H ITEM DIFFICULTY INDICES//(10F7.3))
0204            5  FORMAT (    8H SUBTEST,5X,5HITEMS///)
0205            6  FORMAT (1X,I5,5X,20I4/(11X,20I4))
```

```
FORTRAN IV G LEVEL  21              MAIN            DATE = 73097        11/11/21              PAGE 0009

 0206          7    FORMAT (//16H REPLICATICN NO.I5,10X,9HPARAMETER,
                   111X,15HPOOLED ESTIMATE,10X,24HESTIMATED SE OF ESTIMATE//
                   231X,4HMEANF31.5,F34.5/
                   331X,3HVARF32.5,F34.5/
                   431X,2HU3F33.5,F34.5/
                   531X,2HU4F33.5,F34.5/
                   631X,6HVAR(I)F29.5,F34.5/
                   731X,5HTHETAF30.5,F34.5/
                   831X,20HRELIABILITY OF TOTALF15.5//)
 0207         11    FORMAT (////45X,42HSUMMARY (RESULTS POOLED OVER REPLICATIONS)//
                   110X,9HPARAMETER,10X,13HMEAN ESTIMATE,10X,16HSE(REPLICATIONS),
                   25X,18HMEAN SE(JACKKNIFE),6X,17HSD(SE(JACKKNIFE))//
                   310X,4HMEANF28.7,F26.7,2F23.7/
                   410X,3HVARF29.7,F26.7,2F23.7/
                   510X,2HU3F30.7,F26.7,2F23.7/
                   610X,2HU4F30.7,F26.7,2F23.7/
                   710X,6HVAR(I)F26.7,F26.7,2F23.7/
                   810X,5HTHETAF27.7,F26.7,2F23.7)
 0208         12    FORMAT (20H ALL INPUT PRCCESSED)
 0209         14    FORMAT (//7H MOMENT,10X,16HTHEORETICAL DIST,10X,14HTRUNCATED DIST
                   2//4H U1'E29.7,E24.7/3H U2E30.7,E24.7/3H U3E30.7,E24.7/3H U4E30.7,
                   2E24.7/)
 0210         18    FORMAT (10X,20HRELIABILITY OF TOTALF12.7,F26.7)
 0211         19    FORMAT (//47H ITEM DIFFICULTY INDICES FOR ALL DATA GENERATED
                   1////(10F7.3))
 0212         20    FORMAT (10X,12HSE(U1')-HLN ,F20.7,F26.7/10X,11HSE(U2)-HLN
                   1F21.7,F26.7)
 0213         25    FORMAT (5X,8E15.5)
 0214               END
```

```
FORTRAN IV G LEVEL  21                MAIN             DATE = 73097         11/11/21            PAGE C010

  *OPTIONS IN EFFECT*  ID,EBCDIC,SOURCE,NOLIST,NODECK,LOAD,NOMAP
  *OPTIONS IN EFFECT*  NAME = MAIN    , LINECNT =       56
  *STATISTICS*    SOURCE STATEMENTS =       214,PROGRAM SIZE =     8240
  *STATISTICS*  NO DIAGNCSTICS GENERATED
```

```
FORTRAN IV G LEVEL  21             ALLOC          DATE = 73097        11/11/21          PAGE 0001

0001              SUBROUTINE ALLOC (NT,IPT,ISAMP,KPOP)
            C
            C  RANDOM ASSIGNMENT OF ITEMS TC SUBTESTS
            C
0002              COMMON X(300),LT(1500)
0003              DIMENSICN L(150)
0004         40   NN=NT*IPT
            C
            C  MODIFICATION OF SAMPLING PLAN IF NN .GT. 1500
            C
0005              IF ( NN .LE. 1500 ) GO TC 60
0006              IF ( IPT .LE. 150 ) GO TO 50
0007              IPT=150
0008              WRITE (6,3) IPT
0009              GO TO 40
0010         50   NT=1500/IPT
0011              WRITE (6,4) NT
0012         60   CONTINUE
            C
            C
0013              K=0
0014              DC 100 I=1,KPOP
0015        100   L(I)=I
0016              IF ( ISAMP .EQ. 1 .AND. NN .GE. KPOP ) GO TO 600
0017        150   IF ( ISAMP-1 ) 200,200,400
            C
            C  UNBALANCED DESIGN  (TK LT/EQ/GT KPOP)
            C
0018        200   CONTINUE
0019              DO 290 I=1,NT
0020              DO 270 J=1,IPT
0021        230   CALI RUNIF(R)
0022              JJ=R*KPOP+1.
0023              IF ( JJ .GT. KPOP ) JJ=KPCP
0024              IF ( L(JJ) .LT. 0 ) GO TC 230
0025              K=K+1
0026              LT(K)=L(JJ)
0027        270   L(JJ)=-L(JJ)
0028              IF ( ISAMP .EQ. 1 ) GO TC 290
0029              DO 280 J=1,KPOP
0030        280   L(J)=IABS(L(J))
0031        290   CONTINUE
0032              RETURN
            C
            C  BALANCED DESIGN  (TK GE KPOP AND TK IS INTEGER MULTIPLE OF KPOP)
            C
0033        400   NMULT=NN/KPOP
0034              IF ( FICAT(NN)/KPCP .NE. FLOAT(NMULT) ) GO TO 500
0035              NSTOP=KPOP-IPT
0036              DO 490 I=1,NMULT
0037              DO 450 J=1,NSTOP
0038        430   CALL RUNIF(R)
0039              JJ=R*KPOP+1.
0040              IF ( JJ .GT. KPOP ) JJ=KPCP
```

```
FORTRAN IV G LEVEL  21                 ALLOC              DATE = 73097          11/11/21                 PAGE 0002

0041                IF ( L(JJ) .LT. 0 ) GO TO 430
0042                K=K+1
0043                LT(K)=L(JJ)
0044          450   L(JJ)=-L(JJ)
0045                DO 460 J=1,KPOP
0046                IF ( L(J) .LT. 0 ) GO TO 460
0047                K=K+1
0048                LT(K)=L(J)
0049          460   CONTINUE
0050                DO 470 J=1,KPOP
0051          470   L(J)=IABS(L(J))
0052          490   CONTINUE
0053                RETURN
             C
             C   IF ISAMP=2, (NT*IPT) MUST BE AN EVEN MULTIPLE OF KPOP. WHEN THIS IS
             C   NOT THE CASE, ISAMP IS RESET.
             C
0054          500   IF ( NN .GT. KPOP ) ISAMP=0
0055                IF ( NN .LT. KPOP ) ISAMP=1
0056                WRITE (6,1) ISAMP
0057                GO TO 150
             C
             C   IF ISAMP=1, (NT*IPT) MUST BE LESS THAN KPOP.  WHEN THIS IS NOT
             C   THE CASE, ISAMP IS RESET.
             C
0058          600   ISAMP=0
0059                IF ( FLOAT(NN)/KPOP .EQ. FIOAT(NN/KPOP) ) ISAMP=2
0060                WRITE (6,2) ISAMP
0061                GO TO 150
             C
             C
0062            1   FORMAT (29H TK NOT INTEGER MULTIPLE OF K//30H ITEM SAMPLING SWITCH
                  2 RESET TOI5//)
0063            2   FORMAT (30H ITEM SAMPLING SWITCH RESET TOI5//)
0064            3   FORMAT (28H TK EXCEEDS 1500, K RESET TO I4//)
0065            4   FORMAT (28H TK EXCEEDS 1500, T RESET TO I4//)
0066                END
```

```
FORTRAN IV G LEVEL  21                 ALLOC            DATE = 73097          11/11/21             PAGE 0003

  *OPTIONS IN EFFECT*  ID,EBCDIC,SOURCE,NOLIST,NODECK,LOAD,NOMAP
  *OPTIONS IN EFFECT*  NAME = ALLOC    , LINECNT =       56
  *STATISTICS*    SOURCE STATEMENTS =        66,PROGRAM SIZE =      2658
  *STATISTICS*  NO DIAGNOSTICS GENERATED
```

```
FORTRAN IV G LEVEL  21              BLK DATA          DATE = 73097          11/11/21              PAGE 0001

0001            BLOCK DATA
0002            INTEGER A,B,C,D,E,F,G,H
0003            COMMON /BLOCK5/ N,AA(90),BE(85),CC(45),DD(60),EE(10)
0004            COMMON /BLOCK8/ A(135),B(135),C(135),D(135),E(21),F(55),
               1     G(55),H(135),I(135),J(135),K(135),L(21)
0005            DATA AA/
               10.2,0.2,0.3,0.3,0.3,0.3,0.3,0.5,0.6,0.6,
               20.6,0.6,0.6,0.8,0.8,0.8,1.0,1.0,1.5,0.0,
               30.0,0.0,0.0,0.0,0.0,0.0,0.1,0.1,0.1,0.1,
               40.1,0.1,0.1,0.2,0.2,0.2,0.2,0.2,0.3,0.3,
               50.4,0.4,0.4,0.4,0.4,0.4,0.5,0.5,0.5,0.5,
               60.5,0.6,0.7,0.7,0.7,0.7,0.7,0.8,0.8,0.9,
               70.9,0.9,0.9,1.0,1.0,1.1,1.1,1.1,1.2,1.2,
               81.2,1.3,1.3,1.4,1.4,1.5,1.6,1.7,1.8,0.0,
               90.4,0.4,0.7,0.9,0.9,0.9,1.1,1.1,1.1,1.1/
0006            DATA BB/
               11.3,1.3,1.3,1.3,1.3,1.3,1.4,1.4,1.6,1.6,
               21.6,1.6,1.6,1.6,1.7,1.7,1.7,1.8,1.9,1.9,
               31.9,1.9,1.9,1.9,1.9,1.9,2.0,2.0,2.0,2.0,
               42.0,2.0,2.0,2.1,2.1,2.1,2.1,2.1,2.1,2.2,
               52.2,2.2,2.2,2.3,2.3,2.3,2.4,2.4,2.5,2.6,
               60.7,1.1,1.3,0.4,1.0,1.9,1.4,0.9,0.8,0.6,
               70.5,1.2,1.6,1.7,0.3,1.5,2.0,1.8,2.2,0.2,
               82.5,2.3,2.4,2.1,0.1,2.7,0.0,2.6,2.8,2.9,
               9.9432165,.9464093,.9494969,.9525784,.9555568/
0007            DATA CC/
               1.9584896,.9613885,.9641983,.9667888,.9693678,
               2.9719366,.9744750,.9769426,.9792129,.9812336,
               3.9832494,.9850208,.9864483,.9878070,.9891104,
               4.9902074,.9912605,.9922363,.9931582,.9940219,
               5.9948456,.9955013,.9958897,.9962684,.9973002,
               6.9422782,.9455721,.9485514,.9511653,.9549863,
               7.9566914,.9604850,.9638041,.9665718,.9689170,
               8.9712917,.9742013,.9761328,.9784229,.9805795,
               9.9830652,.9842241,.9863252,.9871416,.9888329/
0008            DATA DD/
               1.9894908,.9907816,.9917306,.9930633,.9938134,
               2.9942625,.9951108,.9958056,.9960779,.9964138,
               3.973,.996,.992,.920,.998,.982,.990,.996,.985,.959,
               4.942,.994,.986,.985,.890,.988,.980,.983,.977,.843,
               5.973,.975,.974,.978,.755,.970,.501,.971,.968,.967,
               612.5,8.205233 ,6.918654 ,20.0,9.032558 ,
               74.644445 ,6.408631 ,10.0,11.11111  ,14.28571  ,
               816.66667  ,7.510414 ,5.574350 ,5.228862 ,25.0,
               95.964524 ,4.395120 ,4.920813 ,3.963179 ,33.33333  /
0009            DATA EE/
               13.442796 ,3.774884 ,3.602029 ,4.169066 ,50.0,
               23.159251 ,100.0,3.295642 ,3.032490 ,2.914378 /
0010            DATA A/
               1  1, -6, -6,  3,  3,  8,  8, -6, -6, 24,  6,  6,-12,-12,-24,
               2-24,  6, -6, -6,-24,-24, 36, 36, 24, 24, 48,  3, 16,-18,-18,
               3-48,-48, 36,  1,  0, -1,  0, -2,  0,  2,  0, -4, -1, -1,  4,
               4  2,  6,  4, -2,  2,  1,  6,  4,-12, -6, -6, -8,-12, -1, -4,
               5  6,  6, 16, 12,-12,  1,  0, -1,  0, -2,  0,  2, -4, -1, -1,
```

```
FORTRAN IV G LEVEL  21                BLK DATA           DATE = 73097         11/11/21              PAGE 0002

                 6  2,  4,  4,  6, -2,  1,  2,  4,  6, -6,-12, -8, -6,-12, -1,
                 7 -4,  6,  6, 12, 16,-12,  1,  0,  0,  0, -1,  0,  0,  0,  0,
                 8 -4,  0,  0,  0,  2, -2,  0,  0,  0,  6,  0,  0,  8,  0,  1,
                 9  0, -3, -6, -8,  0,  6,  1,  0,  0,  0, -1,  0,  0,  0,  0/
 0011             DATA B/
                 1 -4,  0,  0,  2,  0, -2,  0,  0,  0,  6,  8,  0,  0,  1,  0,
                 2 -6, -3,  0, -8,  6,  1,  0, -1,  0,  0,  0,  0,  0,  0, -3,
                 3  0,  0,  0,  0, -3,  0,  6,  0,  3,  0,  6,  0,  2, -3,  0,
                 4 -8, -6,  6,  1,  0, -1,  0,  0,  0,  0,  0,  0, -3,  0,  0,
                 5  0,  0, -3,  0,  6,  0,  3,  6,  0,  2,  0, -3, -6, -8,  6,
                 6  1,  0,  0,  0,  0,  0,  0,  0,  0,  0,  0,  0,  0,  0, -6,
                 7  0,  0,  0,  0,  0,  0,  3,  0,  8,  0, -6,  1,  0,  0,  0,
                 8  0,  0,  0,  0,  0,  0,  0,  0,  0,  0, -6,  0,  0,  0,  0,
                 9  0,  0,  3,  0,  8, -6,  1,  0,  0, -1, -1, -1, -1,  1,  0/
 0012             DATA C/
                 1  0, -1, -1,  2,  2,  2,  2,  3,  0,  1, -2, -2, -4, -4,  4,
                 2  1,  0,  0,  0, -2, -2,  0, -1, -1,  0,  0,  2,  2,  2,  2,
                 3  4,  1,  0, -2, -2, -4, -4,  4,  1,  0,  0,  0,  0,  0, -1,
                 4 -1, -2, -2,  2,  2,  2,  2,  2,  1,  2, -2, -2, -4, -4,  4,
                 5  1,  0,  0,  0, -1,  0,  0,  0,  0, -1,  0,  0, -2,  0,  0,
                 6  0,  1,  2,  2,  0, -2,  1,  0,  0, -1,  0,  0,  0,  0,  0,
                 7 -1, -2,  0,  0,  0,  0,  2,  1,  0,  2, -2,  1,  0,  0,  0,
                 8  0,  0,  0, -1,  0, -1,  0, -2,  0,  0,  1,  0,  2,  2, -2,
                 9  1,  0,  0,  0,  0,  0,  0, -1,  0, -1, -2,  0,  0,  0,  1/
 0013             DATA D/
                 1  2,  2, -2,  1,  0,  0,  0,  0,  0,  0,  0,  0,  0,  0,  0,
                 2 -1, -1,  0,  0,  1,  1,  0,  0,  0, -1,  0,  0, -2,  0, -1,
                 3  0,  1,  2,  2,  0, -2,  1,  0,  0,  0, -1, -2,  0,  0, -1,
                 4  0,  2,  1,  0,  2, -2,  1,  0, -1,  0, -1,  0, -1,  0, -1,
                 5  1,  0,  2,  2, -2,  1,  0, -1,  0, -1, -1,  0, -1,  0,  1,
                 6  2,  2, -2,  1,  0,  0,  0,  0,  0,  0, -1,  0, -2,  0,  2,
                 7  1,  0,  0,  0,  0,  0,  0, -1,  0, -2,  2,  1,  0,  0,  0,
                 8  0, -1,  0,  0, -1,  1,  1,  0,  0,  0,  0, -1, -1,  0,  1,
                 9  1,  0,  0,  0,  0, -1, -1,  1,  1,  0, -1, -1,  0,  0,  1/
 0014             DATA E/
                 1  1,  0,  0, -1, -1,  1,  1,  0,  0,  0, -1,  1,  0,  0, -1,
                 2  1,  0, -1,  1, -1,  1/
 0015             DATA F/
                 1  1, -3, -3,  2,  2,  6,  3, -6, -6,  4,  1,  0, -1,  0, -2,
                 2 -1,  3,  2, -2,  1,  0, -1, -2, -1,  2,  3, -2,  1,  0,  0,
                 3  0, -3,  0,  2,  1,  0,  0,  0, -3,  2,  1,  0, -1, -1,  1,
                 4  1, -1, -1,  1,  1,  0, -1,  1, -1,  1/
 0016             DATA G/
                 1  1, -1,  1, -1,  0,  1,  2, -3,  0,  1,
                 2  2,  0, -3,  0,  1,  1, -1, -1,  0,  0,
                 3  1,  1, -1, -1,  0,  0,  0,  1, -2,  3,
                 4  2, -1,  0, -2, -1,  1, -2,  2,  3,  0,
                 5 -1, -2, -1,  0,  1,  4, -6, -6,  2,  2,
                 6  6,  3, -3, -3,  1/
 0017             DATA H/
                 1  1, -1,  1, -1,  0,  1,  1, -2,  0,  1,  1,  0, -2,  0,  1,
                 2  2, -3,  0,  0,  0,  1,  2,  0, -3,  0,  0,  0,  1, -6, 12,
                 3  0, -3,  0, -4,  0,  1, -6,  0, 12,  0, -3,  0, -4,  0,  1,
                 4  1, -1, -1,  0,  0,  0,  0,  0,  0,  1,  1, -1, -1,  0,  0,
```

```
              5  0,  0,  0,  0,  0,  1,  1, -1, -1,  0,  0,  0,  0,  0,  0,
              6  0,  0,  1, -1,  2,  1, -1,  0,  0,  0, -2,  0, -2,  0,  0,
              7  1, -1,  1,  2,  0, -1,  0,  0,  0,  0, -2,  0,  0,  0,  1,
              8 -2,  3,  2,  0,  0, -1,  0,  0,  0, -2, -1,  0,  0,  0,  1,
              9 -2,  2,  3,  0,  0,  0, -1,  0,  0, -2, -1,  0,  0,  0,  0/
0018           DATA I/
              1  1,  1, -2, -2,  1,  1,  0,  0,  0,  0,  4,  0,  0, -2, -2,
              2  0,  0,  1, -1,  2,  1, -1,  0,  0,  0,  0,  0,  0, -1, -1,
              3  0,  0,  0,  0,  0,  1, -1,  1,  2,  0, -1,  0,  0,  0,  0,
              4  0, -1, -1,  0,  0,  0,  0,  0,  0,  1, -2,  3,  2,  0,  0,
              5 -1,  0,  0,  0, -2,  0, -1,  0,  0,  0,  0,  0,  0,  0,  1,
              6 -2,  2,  3,  0,  0,  0, -1,  0,  0, -2,  0, -1,  0,  0,  0,
              7  0,  0,  0,  0,  0,  1,  6,-12, -6,  3,  0,  4,  0, -1,  0,
              8  8,  2,  2, -2,  0, -2,  0,  0, -1,  0, -2,  0,  1,  6, -6,
              9-12,  0,  3,  0,  4,  0, -1,  8,  2,  2,  0, -2,  0, -2,  0/
0019           DATA J/
              1  0, -1,  0, -2,  0,  1,  2, -3, -4,  0,  2,  1,  0,  0,  0,
              2  4,  1,  1,  0, -2, -1,  0,  0,  0, -1, -1,  0,  0,  0,  1,
              3  2, -4, -3,  2,  0,  0,  1,  0,  0,  4,  1,  1, -2,  0,  0,
              4 -1,  0, -1,  0,  0, -1,  0,  0,  0,  1,  4, -6, -6,  0,  0,
              5  2,  2,  0,  0,  6,  2,  1,  0,  0, -2, -2,  0,  0,  0, -1,
              6 -1,  0,  0,  0,  0,  1,  1, -2, -2,  1,  1,  0,  0,  0,  0,
              7  0,  2,  2,  0,  0,  0,  0,  0, -2, -2,  0,  0,  0,  0,  0,
              8  0,  0,  1,  4, -6, -6,  0,  0,  2,  2,  0,  0,  6,  0,  3,
              9  0,  0,  0,  0,  0,  0,  0, -3, -3,  0,  0,  0,  0,  0,  0/
0020           DATA K/
              1  1, -6, 12, 12, -3, -6, -4,  0,  1,  0,-16, -4, -4,  4,  8,
              2  4,  0, -2,  2,  4,  4,  0, -2,  0, -4,  0,  0, -1,  0,  1,
              3 -6, 12, 12, -6, -3,  0, -4,  0,  1,-16, -4, -4,  8,  4,  0,
              4  4, -2,  4,  2,  0,  4,  0, -2,  0, -4,  0, -1,  0,  0,  1,
              5-12, 24, 18, -6,  0, -8, -6,  2,  0,-24, -6, -6,  6,  0,  6,
              6  6,  0,  3,  0,  6,  6, -3,  0,  0, -3, -3,  0, -1,  0,  0,
              7  1,-12, 18, 24,  0, -6, -6, -8,  0,  2,-24, -6, -6,  0,  6,
              8  6,  6,  0,  0,  3,  6,  6,  0, -3, -3,  0, -3,  0, -1,  0,
              9  0,  0,  1, 36,-72,-72, 18, 18, 24, 24, -6, -6, 96, 24, 24/
0021           DATA L/
              1-24,-24,-24,-24,  6,-12,-12,-24,-24, 12, 12, 12, 12, 12,  3,
              2  4, -3, -3, -4, -4,  1/
0022           END
```

```
FORTRAN IV G LEVEL  21                 BLK DATA         DATE = 73097        11/11/21          PAGE 0004

  *OPTIONS IN EFFECT*  ID,EBCDIC,SOURCE,NOLIST,NODECK,LOAD,NOMAP
  *OPTIONS IN EFFECT*  NAME = BLK DATA, LINECNT =       56
  *STATISTICS*  NO DIAGNOSTICS GENERATED
```

```
FORTRAN IV G LEVEL  21                BMOMTS             DATE = 73097          11/11/21                PAGE C001

0001               SUBROUTINE BMOMTS (V1,U2,U3,U4,KPOP,C)
             C
             C  DETERMINATION OF MINIMUM AND MAXIMUM VALUES FOR EACH ESTIMATED
             C  CENTRAL MOMENT
             C
0002               COMMON /BLOCK1/ S(4),NS(2),INTPRT
0003               COMMON /BLOCK6/ SSUM(4),X(10,157),B(6,4),W(6)
0004               DIMENSION C(4,7)
             C
             C  CONVERT CENTRAL MOMENTS TO MOMENTS ABOUT ORIGIN
             C
0005               V2=U2+V1*V1
0006               V3=U3+3.*V1*V2-2.*V1**3
0007               V4=U4+4.*V3*V1-6.*V2*V1*V1+3.*V1**4
0008               SSUM(1)=V1
0009               SSUM(2)=V2
0010               SSUM(3)=V3
0011               SSUM(4)=V4
0012               DO 40 I=1,4
0013               DO 40 J=1,7
0014          40   C(I,J)=0.
0015               DO 50 I=1,4
0016          50   C(I,1)=SSUM(I)
0017               C(1,4)=0.
0018               C(2,4)=U2
0019               C(3,4)=U3
0020               C(4,4)=U4
0021               X(1,5)=7
0022               X(1,6)=KPOP+1
0023               NJ=KPOP+7
             C
             C  COMPUTE BOUNDARIES FOR MOMENTS ABOUT ORIGIN AND CONVERT TO
             C  BOUNDARIES FOR CENTRAL MOMENTS
             C
0024               IF ( INTPRT .EQ. 0 ) GO TO 90
0025               WRITE (6,5)
0026               WRITE (6,1)
0027          90   DO 100 I=4,6
0028               MOM=I-2
0029               CALL SETUP (I,NJ,I,1.)
0030               CALL LE(NERR)
0031               IF ( NERR .EQ. 0 ) GO TO 110
0032               CALL MOMVAL (C(MOM,2),MOM,C(MOM,5))
0033               CALL SETUP (I,NJ,I,-1.)
0034               CALL LE (NERR)
0035               CALL MOMVAL (C(MOM,3),MOM,C(MOM,6))
0036               C(MOM,7)=0.
0037               IF ( C(MOM,1) .LT. C(MOM,2) .OR. C(MOM,1) .GT. C(MOM,3) )GO TO 110
0038         100   CONTINUE
0039               GO TO 115
0040         110   C(MOM,7)=1.
             C
             C
0041         115   IF ( C(3,5) .LE. C(3,6) ) GO TO 130
```

```
FORTRAN IV G LEVEL  21                 BMOMTS          DATE = 73097        11/11/21              PAGE 0002

                  TEMP=C(3,5)
                  C(3,5)=C(3,6)
                  C(3,6)=TEMP
             130  CONTINUE
            C
            C  PRINTCUT EOUNDARIES FOR MOMENTS AND APPRAISE ESTIMATED MOMENTS
            C
                  IF ( INTPRT .EQ. 0 ) GO TO 155
                  I=1
                  WRITE (6,4) I,C(1,1),C(1,4)
                  DO 150 I=2,4
                  IF ( C(I,1) .LT. C(I,2) .CE. C(I,1) .GT. C(I,3) ) GO TO 145
                  WRITE (6,2) I,(C(I,J),J=1,6)
                  GO TO 150
             145  WRITE (6,3) I,(C(I,J),J=1,6)
             150  CONTINUE
            C
            C  FIXUP BAD MOMENT ESTIMATES BY CHANGING ESTIMATED BAD MOMENT
            C  TO MEAN OF ESTIMATED BOUNDARIES
            C
                  WRITE (6,5)
             155  IF ( C(2,7) .EQ. 0. ) GO TC 160
                  V2=(C(2,2)+C(2,3))/2.
                  SSUM(2)=V2
                  C(2,1)=V2
                  C(2,4)=V2-V1*V1
                  IF ( INTPRT .EC. 1 ) WRITE (6,6)
                  GO TC 90
             160  IF ( C(3,7) .EQ. 0. ) GO TO 170
                  V3=(C(3,2)+C(3,3))/2.
                  SSUM(3)=V3
                  C(3,1)=V3
                  C(3,4)=V3-3.*V1*V2+2.*V1**3
                  IF ( INTPRT .EQ. 1 ) WRITE (6,6)
                  GO TC 90
             170  IF ( C(4,7) .EQ. 0. ) GO TC 200
                  C(4,1)=(C(4,2)+C(4,3))/2.
                  C(4,4)=(C(4,5)+C(4,6))/2.
                  IF ( INTPRT .EQ. 0 ) GO TC 200
                  WRITE (6,6)
                  I=4
                  WRITE (6,4) I,C(4,1),C(4,4)
             200  U2=C(2,4)
                  U3=C(3,4)
                  U4=C(4,4)
                  RETURN
               1  FORMAT (60H MINIMUM AND MAXIMUM POSSIBLE VALUES FOR ESTIMATED MOME
                 1NTS  //23X,12HABCUT ORIGIN,35X,10HABOUT MEAN/
                 210X,38H--------------------------------------,
                 3 7X,38H--------------------------------------//
                 41X,6HMOMENT,3X,38HESTIMATE        MINIMUM        MAXIMUM,
                 57X,           38EESTIMATE        MINIMUM        MAXIMUM
                 69X,9HAPPRAISAL//)
               2  FORMAT (1X,I3,3X,6E15.7,8X,2HOK)
```

```
FORTRAN IV G LEVEL  21                BMOMTS            DATE = 73097         11/11/21            PAGE 0003

 0083          3     FORMAT (1X,I3,3X,6E15.7,8X,6HNOT OK)
 0084          4     FORMAT (1X,I3,3X,E15.7,30X,E15.7)
 0085          5     FORMAT (///)
 0086          6     FORMAT (   31H ATTEMPTED FIXUP OF BAD MOMENTS///)
 0087                END
```

```
FORTRAN IV G LEVEL  21               BMOMTS          DATE = 73097        11/11/21           PAGE C004

  *OPTIONS IN EFFECT*  ID,EBCDIC,SOURCE,NOLIST,NODECK,LOAD,NOMAP
  *OPTIONS IN EFFECT*  NAME = BMOMTS  , LINECNT =       56
  *STATISTICS*    SOURCE STATEMENTS =       87,PROGRAM SIZE =     2816
  *STATISTICS*  NO DIAGNOSTICS GENERATED
```

```
FORTRAN IV G LEVEL  21              DATA            DATE = 73097         11/11/21            PAGE 0001

0001             SUBROUTINE DATA (X)
           C
           C  GENERATION OF ITEM SCORES AND TEST SCORE FOR HYPOTHETICAL EXAMINEE
           C
0002             REAL MPOP
0003             INTEGER TSCORE
0004             COMMON /BLOCK1/ YBAR,YSD,MPOP,SPOP,KPOP,NDIST,INTPRT
0005             COMMON /BLOCK2/ RND(150)
0006             COMMON /BLOCK3/ P(150),Q(150),NSUB
0007             DIMENSION X(150)
           C
           C  GENERATE TOTAL TEST SCORE
           C
0008             NSUB=NSUB+1
0009             CALL RANDND (Z)
0010             GO TO (215,220,225), NDIST
0011         215 TEMP=Z*SPOP+MPOP
0012             GO TO 230
0013         220 TEMP=EXP(Z*YSD+YBAR)
0014             GO TO 230
0015         225 TEMP=KPOP-EXP(Z*YSD+YBAR)
0016         230 IF ( TEMP .LE. 0. ) GO TO 235
0017             IF ( TEMP .GE. FLOAT(KPOP) ) GO TO 238
0018             TSCORE=TEMP
0019             KSUB=TSCORE+1
0020             IF ( TEMP .GT. RND(KSUB) ) TSCORE=KSUB
0021             GO TO 239
0022         235 TSCORE=0
0023             GO TO 239
0024         238 TSCORE=KPOP
0025         239 CONTINUE
           C
           C  GENERATE ITEM SCORES FOR EXAMINEE
           C
0026             DO 240 J=1,KPOP
0027         240 X(J)=0.
0028             IF ( TSCORE .EQ. 0 ) GO TO 300
0029             IF ( TSCORE .LT. KPOP ) GO TO 248
0030             DO 242 J=1,KPOP
0031         242 X(J)=1.
0032             GO TO 300
0033         248 KOUNT=0
0034             DO 250 J=1,KPOP
0035             IF ( Q(J) .GT. P(J) ) GO TO 250
0036             KOUNT=KOUNT+1
0037             IF ( KOUNT .GT. TSCORE ) GO TO 300
0038             X(J)=1.
0039         250 CONTINUE
0040             DO 260 J=1,KPOP
0041             IF ( X(J) .EQ. 1. ) GO TO 260
0042             KOUNT=KOUNT+1
0043             IF ( KOUNT .GT. TSCORE ) GO TO 300
0044             X(J)=1.
0045         260 CONTINUE
```

```
FORTRAN IV G LEVEL  21                DATA            DATE = 73097        11/11/21            PAGE 0002

      300  DO 320 J=1,KPCP
      320  Q(J)=(Q(J)*(NSUB-1.)+X(J))/NSUB
           RETURN
           END
```

```
FORTRAN IV G LEVEL  21              DATA              DATE = 73097         11/11/21            PAGE 0003

  *OPTIONS IN EFFECT*  ID,EBCDIC,SOURCE,NOLIST,NODECK,LOAD,NOMAP
  *OPTIONS IN EFFECT*  NAME = DATA      LINECNT =       56
  *STATISTICS*    SOURCE STATEMENTS =       49,PROGRAM SIZE =     1354
  *STATISTICS*  NO DIAGNOSTICS GENERATED
```

```
FORTRAN IV G LEVEL  21           ESTORS           DATE = 73097        11/11/21            PAGE C001

                SUBFCUTINE ESTORS (XX,NSPT,IPT,RSLT,VARU1,VARU2)
          C
          C  ESTIMATION OF PARAMETERS FBOM INDIVIDUAL SUBTEST RESULTS
          C
                INTEGER FS,AS,CAS,CFS
                COMMON /BLOCK1/ DUMMY(4),KPOP,NDIST,INTPRT
                COMMCN /BLOCK8/ FS(561),AS(55),CAS(55),CFS(561)
                DIMENSION XIP(100),YIP(100),ZIP(100),UIP(100),VIP(100),
               2     XPJ(150),YPJ(150),ZPJ(150),UPJ(150),VPJ(150),
               3     XX(100,150),RSLT(8),
               4     S2(4),S3(10),S4(33),C2(4),C3(10),C4(33),
               5     D(4),A(10),F(33),DD(4),AA(10),FF(33)
          C
          C
                DO 70 I=1,IPT
                XPJ(I)=0.
                YPJ(I)=0.
                ZPJ(I)=0.
                UPJ(I)=0.
             70 VPJ(I)=0.
                DO 85 I=1,33
             85 S4(I)=0.
          C
          C
                DO 100 I=1,NSPT
                XIP(I)=0.
                YIP(I)=0.
                ZIP(I)=0.
                UIP(I)=0.
                VIP(I)=0.
                DO 100 J=1,IPT
                TEMP=XX(I,J)
                XIP(I)=XIP(I)+TEMP
                XPJ(J)=XPJ(J)+TEMP
                YIP(I)=YIP(I)+TEMP**2
                YPJ(J)=YPJ(J)+TEMP**2
                ZIP(I)=ZIP(I)+TEMP**3
            100 ZPJ(J)=ZPJ(J)+TEMP**3
                Q2=0.
                Q4=0.
                Q7=0.
                DO 120 I=1,IPT
                TEMP=XPJ(I)
                Q2=Q2+TEMP**2
                Q4=Q4+TEMP**3
                Q7=Q7+YPJ(I)*XPJ(I)
                S4(9)=S4(9)+TEMP**4
                S4(30)=S4(30)+YPJ(I)**2
            120 S4(32)=S4(32)+XPJ(I)*ZPJ(I)
                Q1=0.
                Q3=0.
                Q6=C.
                DO 130 I=1,NSPT
                TEMP=XIP(I)
```

```
          Q1=Q1+TEMP**2
          Q3=Q3+TEMP**3
          Q6=Q6+YIP(I)*XIP(I)
          S4(8)=S4(8)+TEMP**4
          S4(29)=S4(29)+YIP(I)**2
          S4(31)=S4(31)+XIP(I)*ZIP(I)
    130   S4(22)=S4(22)+YIP(I)*TEMP**2
          Q8=0.
          S3(6)=0.
          S3(10)=0.
          DO 145 I=1,NSPT
          DO 140 J=1,IPT
          TEMP=XX(I,J)*XIP(I)
          UIP(I)=UIP(I)+TEMP
          UPJ(J)=UPJ(J)+TEMP
          TEMP=XX(I,J)*XPJ(J)
          VIP(I)=VIP(I)+TEMP
          VPJ(J)=VPJ(J)+TEMP
          Q8=Q8+XX(I,J)**4
          S3(10)=S3(10)+XX(I,J)**3
          TEMP=XX(I,J)*XIP(I)*XPJ(J)
          S3(6)=S3(6)+TEMP
    140   S4(26)=S4(26)+XX(I,J)*TEMP
          S4(14)=S4(14)+VIP(I)**2
          S4(24)=S4(24)+YIP(I)*VIP(I)
    145   S4(15)=S4(15)+UIP(I)*VIP(I)
          Q5=0.
          XPP=0.
          YPP=0.
          ZPP=0.
          DO 150 I=1,IPT
          XPP=XPP+XPJ(I)
          YPP=YPP+YPJ(I)
          ZPP=ZPP+ZPJ(I)
          Q5=Q5+UPJ(I)*XPJ(I)
          S4(13)=S4(13)+UPJ(I)**2
          S4(16)=S4(16)+UPJ(I)*VPJ(I)
          S4(23)=S4(23)+YPJ(I)*XPJ(I)**2
    150   S4(25)=S4(25)+YPJ(I)*UPJ(I)
          Q9=0.
          NSTOP=IPT-1
          DO 160 I=1,NSTOP
          NSTART=I+1
          DO 160 J=NSTART,IPT
          TEMP=0.
          DO 155 K=1,NSPT
    155   TEMP=TEMP+XX(K,I)*XX(K,J)
    160   Q9=Q9+2.*TEMP*TEMP
          DO 165 I=1,IPT
    165   Q9=Q9+YPJ(I)**2
C
C  CALCULATION OF SIGMAS REQUIRED IN FIRST, SECOND, THIRD AND FOURTH
C  DEGREE GSM'S
C
```

```
FORTRAN IV G LEVEL  21            ESTORS          DATE = 73097        11/11/21          PAGE 0003

0094              S4(1)=XPP**4
0095              S4(2)=XPP*XPP*Q1
0096              S4(3)=XPP*XPP*Q2
0097              S4(4)=Q1*Q1
0098              S4(5)=Q2*Q2
0099              S4(6)=XPP*Q3
0100              S4(7)=XPP*Q4
0101              S4(10)=XPP*Q5
0102              S4(11)=Q1*Q2
0103              S4(12)=XPP*XPP*YPP
0104              S4(17)=Q9
0105              S4(18)=YPP*Q1
0106              S4(19)=YPP*Q2
0107              S4(20)=XPP*Q6
0108              S4(21)=XPP*Q7
0109              S4(27)=YPP*YPP
0110              S4(28)=XPP*ZPP
0111              S4(33)=Q8
            C
            C
0112              S3(1)=XPP**3
0113              S3(2)=XPP*Q1
0114              S3(3)=XPP*Q2
0115              S3(4)=Q3
0116              S3(5)=Q4
0117              S3(7)=YPP*XPP
0118              S3(8)=Q6
0119              S3(9)=Q7
            C
            C
0120              S2(1)=XPP**2
0121              S2(2)=Q1
0122              S2(3)=Q2
0123              S2(4)=YPP
            C
            C
0124              S1=XPP
            C
            C   CALCULATION OF CONSTANTS ASSOCIATED WITH GSM'S
            C
0125              RC=NSPT*IPT
0126              R1=NSPT-1.
0127              R2=NSPT-2.
0128              R3=NSPT-3.
0129              A1=IPT-1.
0130              A2=IPT-2.
0131              A3=IPT-3.
            C
            C   CONSTANT FOR FIRST DEGREE GSM
            C
0132              C1=1./RC
            C
            C   CONSTANTS FOR SECOND DEGREE GSM'S
            C
```

```
FORTRAN IV G LEVEL  21              ESTORS            DATE = 73097          11/11/21              PAGE 0004

 0133               C2(1)=1./(RC*R1*A1)
 0134               C2(2)=1./(RC*A1)
 0135               C2(3)=1./(RC*R1)
 0136               C2(4)=1./RC
              C
              C
 0137               T=RC*R1*A1
 0138               U=1./(T*R2*R3*A2*A3)
 0139               B=1./(T*R2*A2*A3)
 0140               C=1./(T*R2*R3*A2)
 0141               V=1./(T*A2*A3)
 0142               E=1./(T*R2*R3)
 0143               G=1./(T*R2*A2)
 0144               H=1./(T*A2)
 0145               P=1./(T*R2)
 0146               Q=1./T
              C
              C  CONSTANTS FOR THIRD DEGREE GMS'S
              C
 0147               C3(1)=G
 0148               C3(2)=H
 0149               C3(3)=P
 0150               C3(4)=1./(RC*A1*A2)
 0151               C3(5)=1./(RC*R1*R2)
 0152               C3(6)=Q
 0153               C3(7)=Q
 0154               C3(8)=1./(RC*A1)
 0155               C3(9)=1./(RC*R1)
 0156               C3(10)=1./RC
              C
              C  CONSTANTS FOR FOURTH DEGREE GSM'S
              C
 0157               C4(1)=U
 0158               C4(2)=B
 0159               C4(3)=C
 0160               C4(4)=V
 0161               C4(5)=E
 0162               C4(6)=V
 0163               C4(7)=E
 0164               C4(8)=1./(RC*A1*A2*A3)
 0165               C4(9)=1./(RC*R1*R2*R3)
 0166               C4(10)=G
 0167               C4(11)=G
 0168               C4(12)=G
 0169               C4(13)=H
 0170               C4(14)=P
 0171               C4(15)=H
 0172               C4(16)=P
 0173               C4(17)=Q
 0174               C4(18)=H
 0175               C4(19)=P
 0176               C4(20)=H
 0177               C4(21)=P
 0178               C4(22)=1./(RC*A1*A2)
```

```
FORTRAN IV G LEVEL  21                ESTORS              DATE = 73097           11/11/21              PAGE C005

                C4(23)=1./(RC*R1*R2)
                DO 180 I=24,28
          180   C4(I)=Q
                C4(29)=1./(RC*A1)
                C4(30)=1./(RC*R1)
                C4(31)=1./(RC*A1)
                C4(32)=1./(RC*R1)
                C4(33)=1./RC
        C
        C  CALCULATION OF FIRST DEGREE GSM (FIRST DEGREE BIPOLYKAY)
        C
                Z1=S1*C1
        C
        C  CALCULATION OF SECOND DEGREE GSM'S
        C
                D(1)=(S2(1)-S2(2)-S2(3)+S2(4))*C2(1)
                D(2)=(S2(2)-S2(4))*C2(2)
                D(3)=(S2(3)-S2(4))*C2(3)
                D(4)=S2(4)*C2(4)
        C
        C  CALCULATION OF SECOND DEGREE BIPOLYAKYS
        C
                DD(1)=D(1)
                DD(2)=-D(1)+D(2)
                DD(3)=-D(1)+D(3)
                DD(4)=D(1)-D(2)-D(3)+D(4)
        C
        C  CALCULATION OF THIRD DEGREE GSM'S
        C
                DO 200 J=1,10
                A(J)=0.
                DO 210 I=J,10
                LSUB=(J-1)*10+I-J*(J-1)/2
          210   A(J)=A(J)+AS(LSUB)*S3(I)
          200   A(J)=A(J)*C3(J)
        C
        C  CALCULATION OF THIRD DEGREE BIPOLYKAYS
        C
                DO 220 J=1,10
                AA(J)=0.
                DO 220 I=1,J
                LSUB=(J-1)*10+I-(-20+21*J-J*J)/2
          220   AA(J)=AA(J)+CAS(LSUB)*A(I)
        C
        C  CALCULATION OF FOURTH DEGREE GSM'S
        C
                DO 230 J=1,33
                F(J)=0.
                DO 240 I=J,33
                LSUB=(J-1)*33+I-J*(J-1)/2
          240   F(J)=F(J)+FS(LSUB)*S4(I)
          230   F(J)=F(J)*C4(J)
        C  CALCULATION OF FOURTH DEGREE BIPOLYKAYS
        C
```

```
FORTRAN IV G LEVEL  21                 ESTORS              DATE = 73097            11/11/21                PAGE 0006

0213                DO 250 J=1,33
0214                FF(J)=0.
0215                DO 250 I=1,J
0216                LSUB=(J-1)*33+I-(-65+67*J-J*J)/2
0217           250  FF(J)=FF(J)+CFS(LSUB)*F(I)
              C
              C
              C  ESTIMATION OF CENTRAL MOMENTS OF UNIVERSE SCORES WHEN POPULATION
              C  IS INFINITE
              C
              C  MEAN UNIVERSE SCORE
              C
0218                EU1=Z1*KPOP
              C
              C  SECOND CENTRAL MOMENT
              C
0219                EU2=DD(1)*KPOP+DD(2)*KPOP**2
              C
              C  THIRD CENTRAL MOMENT
              C
0220                EU3=AA(4)*KPOP**3+3.*AA(8)*KPOP**2+KPOP*AA(10)
              C
              C  FOURTH CENTRAL MOMENT
              C
0221                EU4=KPOP**4*(3.*FF(4)+FF(8))+KPOP**3*(12.*FF(13)+6.*FF(18)
                   1   +2.*FF(22))+KPOP**2*(6.*FF(17)+12.*FF(25)+3.*FF(27)
                   2   +4.*FF(31))+KPOP*(3.*FF(30)+FF(33))
              C
              C  CALCULATION OF COMPONENTS OF VARIANCE FOR SUBTEST MATRIX AND
              C  THETA (A FUNCTION OF THE MEAN ITEM RELIABILITY)
              C
0222                CXX=XPP*XPP*C2(4)
0223                SSE=Q1/IPT-CXX
0224                SSI=Q2/NSPT-CXX
0225                SSIE=XPP-CXX-SSE-SSI
0226                ERR=SSIE/(A1*R1)
0227                RSLT(5)=(SSI/A1-ERR)/NSPT
0228                RSLT(6)=(SSE/R1-ERR)/IPT
0229                RSLT(7)=ERR
0230                FMP=(R1*A1-2.)/(R1*A1)
0231                RSLT(8)=(SSE/R1-FMP*ERR)/(IPT*FMP*ERR)
              C
              C
0232                RSLT(1)=EU1
0233                RSLT(2)=EU2
0234                RSLT(3)=EU3
0235                RSLT(4)=EU4
              C
              C  ESTIMATION OF VARIANCE OF SECOND CENTRAL MOMENT
              C
0236                VARU2=(1./NSPT)*(FF(8)+2.*FF(22)/KPO2+FF(29)/KPOP**2)+
                   1 (1./IPT-1./KPOP)*(4.*FF(13)+4.*FF(25)/KPOP+FF(30)/KPOP**2)+
                   2 (2./(NSPT-1.))*(FF(4)+2.*FF(13)/KPOP+FF(17)/KPOP**2)
                   3 +(2./KPOP**2)*(1./(IPT-1.)-1./(KPOP-1.))*FF(17)+(1.-2./KPOP)*
```

```
FORTRAN IV G LEVEL  21              ESTORS           DATE = 73097        11/11/21             PAGE 0007
                       4 (2./(IPT*(IPT-1.))-2./(KPCP*(KPOP-1.)))*FF(17)+(2./(IPT*(NSPT-1.)
                       5  ))*(2.*(FF(13)+FF(18))+4.*FF(25)/KPOP+FF(30)/KPOP**2)+(2./(KPOP
                       6  **2*NSPT*(IPT-1.)))*FF(29)+(1./(NSPT*IPT))*(4.*FF(22)+4.*FF(31)/
                       7 KPOE+FF(33)/KPOP**2)+(2./(KPOP**2*(NSPT-1.)*(IPT-1.)))*(FF(17)+
                       8 FF(27))+(1.-2./KPOP)*(2./(IPT*NSPT*(IPT-1.)))*FF(29)+(1.-2./KPOP)
                       9 *(2./(IPT*(NSPT-1.)*(IPT-1.)))*(FF(17)+FF(27))
 0237                     VARU2 =VARU2 *KPOE**4
                   C
                   C  ESTIMATION OF VARIANCE OF MEAN
                   C
                   C
 0238                     VARU1=DD(4)*C1+DD(2)/NSPT+(1./IPT-1./KPOP)*DD(3)
 0239                     VARU1=VARU1*KEOP**2
                   C
                   C
 0240                     RETURN
                   C
                   C
 0241                     END
```

```
FORTRAN IV G LEVEL  21                 ESTORS            DATE = 73097          11/11/21              PAGE 0008

  *OPTIONS IN EFFECT*  ID,EBCDIC,SOURCE,NOLIST,NODECK,LOAD,NOMAP
  *OPTIONS IN EFFECT*  NAME = ESTORS  , LINECNT =        56
  *STATISTICS*    SOURCE STATEMENTS =       241,PROGRAM SIZE =     13224
  *STATISTICS*  NO DIAGNOSTICS GENERATED
```

```
FORTRAN IV G LEVEL  21                FREAKS            DATE = 73097          11/11/21              PAGE 0001

0001               SUEROUTINE FREAKS (X)
             C
             C  TABULATION OF FREQUENCIES OF EOOLED ESTIMATES
             C
             C
             C  NOTE:   CFQ(1,I) = MINIMUM EOSSIBLE VALUE
             C          CFQ(2,I) = MAXIMUM EOSSIBLE VALUE
             C          CFQ(3,I) = STEP SIZE
             C          CFQ(4,I) = RANGE
             C          CFQ(5,I) = NUMBER OF INTERVALS (STEPS)
             C
0002               INTEGER FQ
0003               COMMON /BLOCK7/ FQ(52,13),CFQ(5,13),NFREQ
0004               DIMENSICN X(13)
             C
             C
0005               DO 100 I=1,13
0006               T=X(I)
0007               NSTEP=CFQ(5,I)
0C08               FMAX=CFQ(2,I)
0009               FMIN=CFQ(1,I)
0010               IF ( T .GE. FMIN .AND. T .LE. FMAX ) GO TO 50
0011               IF ( T .GT. FMAX ) FQ( 52,I)=FQ( 52,I)+1
0012               IF ( T .LT. FMIN ) FQ( 51,I)=FQ( 51,I)+1
0013               GC TC 100
0014            50 IF ( FMIN .GE. 0. ) GO TO 60
0015               R=SQRT(FMIN*FMIN)+T
0016               GO TO 70
0017            60 R=T-FMIN
0018            70 NROW=(R/CFQ(4,I))*CFQ(5,I)+1.
0C19               IF ( NROW .GT. NSTEP ) NECW=NSTEP
0020               FQ(NROW,I)=FQ(NROW,I)+1
0021           100 CONTINUE
             C
             C
0022               RETURN
             C
             C
0023               END
```

```
FORTRAN IV G LEVEL  21              FREAKS          DATE = 73097        11/11/21          PAGE C002

  *OPTIONS IN EFFECT*  ID,EBCDIC,SOURCE,NOLIST,NODECK,LOAD,NOMAP
  *OPTIONS IN EFFECT*  NAME = FREAKS  , LINECNT =       56
  *STATISTICS*    SOURCE STATEMENTS =       23,PROGRAM SIZE =      824
  *STATISTICS*  NO DIAGNOSTICS GENERATED
```

```
FORTRAN IV G LEVEL  21              PRKOUT          DATE = 73097        11/11/21          PAGE 0001

            SUBROUTINE PRKOUT (NREPS)
      C
      C  PRINTOUT OF FREQUENCY DISTRIBUTIONS OF POOLED ESTIMATES
      C
            INTEGER FQ
            REAL MIN,L1,L2
            COMMON /BLOCK7/ FQ(52,13),CFQ(5,13),NFREQ
            DIMENSION LABEL(13)
            DATA LABEL /4HMEAN,4H U2 ,4H U3 ,4H U4 ,
           14HV-P ,4HT'  ,4HV-IE,4HV-E ,
           24HA20 ,4HT-U1,4HT-U2,4HT-U3,4HT-U4/
      C
            FN=NREPS
      C
            DO 100 I=1,13
      C
            WRITE (6,1) LABEL(I)
      C
            MIN=CFQ(1,I)
            STEP=CFQ(3,I)
            NSTEP=CFQ(5,I)
      C
            IF ( NSTEP .GT. 1 .AND. NSTEP .LE.  50 ) GO TO 50
            WRITE (6,2) NSTEP
            GO TO 100
      C
         50 WRITE (6,3)
            IF ( FQ( 51,I) .EQ. 0 ) GO TO 60
            RF=FLOAT(FQ( 51,I))/FN
            WRITE (6,4) FQ( 51,I),RF
         60 DO 70 J=1,NSTEP
            L1=MIN+(J-1)*STEP
            L2=L1+STEP
            RF=0.
            IF ( FQ(J,I) .GT. 0 ) RF=FLOAT(FQ(J,I))/FN
            WRITE (6,5) L1,L2,FQ(J,I),RF
         70 CONTINUE
            IF ( FQ( 52,I) .EQ. 0 ) GO TO 100
            RF=FLOAT(FQ( 52,I))/FN
            WRITE (6,6) FQ( 52,I),RF
        100 CONTINUE
      C
      C
            RETURN
      C
      C
          1 FORMAT (39H1 DISTRIBUTION FOR POOLED ESTIMATES OF ,A4//)
          2 FORMAT (45H NO. OF INTERVALS NOT WITHIN PROGRAM LIMITS (I10,2H ))
          3 FORMAT (12X,8HINTERVAL,19X,9HFREQUENCY, 6X,10HREL. FREQ.//)
          4 FORMAT (1X,17HLESS THAN MINIMUM,20X,I10,5X,F8.5)
          5 FORMAT (1X,E13.7,6H  TO  ,E13.7,5X,I10,5X,F8.5)
          6 FORMAT (1X,20HGREATER THAN MAXIMUM,17X,I10,5X,F8.5)
      C
      C
```

```
FORTRAN IV G LEVEL  21            FRKOUT          DATE = 73097          11/11/21            PAGE 0002
 0038              END
```

```
FORTRAN IV G LEVEL  21                FRKOUT           DATE = 73097        11/11/21             PAGE C003

  *OPTIONS IN EFFECT*  ID,EBCDIC,SOURCE,NOLIST,NODECK,LOAD,NOMAP
  *OPTIONS IN EFFECT*  NAME = FRKOUT  , LINECNT =       56
  *STATISTICS*    SOURCE STATEMENTS =       38,PROGRAM SIZE =     1416
  *STATISTICS*  NO DIAGNOSTICS GENERATED
```

```
FORTRAN IV G LEVEL  21              FRV            DATE = 73097        11/11/21             PAGE 0001

 0001                 FUNCTICN FEV  (R,VV)
               C
               C   EVALUATION OF COS(X)**R*EXP(-V*X)DX FROM -PI/2 TO PI/2 USING
               C   SIMPSON'S RULE
               C
 0002                 V=ABS(VV)
 0003                 PI=3.141593
 0004                 D=PI/100.
 0005                 FRV=0.
 0006                 DO 100 I=1,99,2
 0007                 T=-1.570796+I*D
 0008            100  FRV=FRV+2.*COS(T)**R*EXP(-V*T)+COS(T+D)**R*EXF(-VT-VD)
 0009                 FRV=2.*D*FRV/3.
 0010                 RETUEN
 0011                 END
```

```
FORTRAN IV G LEVEL  21              FRV              DATE = 73097        11/11/21            PAGE 0002

  *OPTIONS IN EFFECT*  ID,EBCDIC,SOURCE,NOLIST,NODECK,LOAD,NOMAP
  *OPTIONS IN EFFECT*  NAME = FRV      , LINECNT =       56
  *STATISTICS*    SOURCE STATEMENTS =        11,PROGRAM SIZE =       736
  *STATISTICS*  NO DIAGNOSTICS GENERATED
```

```
FORTRAN IV G LEVEL  21          GAMMA          DATE = 73097          11/11/21          PAGE 0001

              FUNCTION GAMMA (XX)
      C
      C  COMPUTATION OF GAMMA FUNCTION USING THE RECURSION RELATION AND
      C  POLYNOMIAL APPROXIMATION
      C
              IF ( XX-57. ) 6,6,4
        4     WRITE (6,2)
              GAMMA=1.E75
              RETURN
        6     X=XX
              ERR=1.E-6
              GX=1.
              IF ( X-2. ) 50,50,15
       10     IF ( X-2. ) 110,110,15
       15     X=X-1.
              GX=GX*X
              GO TO 10
       50     IF ( X-1. ) 60,120,110
      C
      C  TEST FOR X NEAR NEGATIVE INTEGER OR ZERO
      C
       60     IF ( X-ERR ) 62,62,80
       62     Y=FLOAT(INT(X))-X
              IF ( ABS(Y)-ERR ) 130,130,64
       64     IF ( 1.-Y-ERR ) 130,130,70
      C
      C  X NOT NEAR A NEGATIVE INTEGER OR ZERO
      C
       70     IF ( X-1. ) 80,80,110
       80     GX=GX/X
              X=X+1.
              GO TO 70
      110     Y=X-1.
              GY=1.+Y*(-.5771017+Y*(.9858540+Y*(-.8764218+Y*(.8328212+Y*
             1(-.5684729+Y*(.2548205+Y*(-.05149930)))))))
              GAMMA=GX*GY
      120     RETURN
      130     WRITE (6,1)
              RETURN
        1     FORMAT (/53H GAMMA ARG WITHIN .000001 OF BEING A NEGATIVE INTEGER/
             1)
        2     FORMAT (/44H GAMMA ARG GT 57, OVERFLOW, ANS SET TO 1.E75/)
              END
```

```
FORTRAN IV G LEVEL  21              GAMMA             DATE = 73097          11/11/21            PAGE 0002

 *OPTIONS IN EFFECT*  ID,EBCDIC,SOURCE,NOLIST,NODECK,LOAD,NOMAP
 *OPTIONS IN EFFECT*  NAME = GAMMA    , LINECNT =       56
 *STATISTICS*    SOURCE STATEMENTS =       31,PROGRAM SIZE =      1036
 *STATISTICS*  NO DIAGNOSTICS GENERATED
```

```
FORTRAN IV G LEVEL  21              LP              DATE = 73097         11/11/21            PAGE 0001

            SUBROUTINE LP (NERR)
      C
      C  LINEAR PROGRAMMING SUBROUTINE
      C
      C  SUBROUTINE WRITTEN ORIGINALLY BY DR. JAMES E. MARTIN (VIRGINIA
      C  POLYTECHNIC INSTITUTE) AND MODIFIED BY DR. RICHARD K. HILL, JR
      C  (VIRGINIA POLYTECHNIC INSTITUTE)
      C
            COMMON /BLOCK6/ SSUM(4),X(10,157),B(6,4),W(6)
            NERR=1
            S=.000001
            NI=X(1,5)
            NJ=X(1,6)+6
            X(2,5)=0.
            DO 30 I=2,NI
         30 X(I,6)=X(I,5)
            DO 32 J=7,NJ
         32 X(2,J)=X(1,J)
            NM=0
            DO 34 LL=1,4
         34 B(2,LL)=X(2,LL)
            DO 50 I=3,NI
            DO 38 J=1,4
         38 B(I,J)=X(I,J)
         50 CONTINUE
            NIT=0
         51 SUMM=0.
            SUMJ=0.
            M=7
            DO 62 J=7,NJ
            DO 52 I=3,NI
            IF ( B(I,2) .NE. X(I,2) ) GO TO 52
            IF ( B(I,3) .NE. X(I,3) ) GO TO 52
            IF ( B(I,4) .NE. X(I,4) ) GO TO 52
            SUMJ=SUMJ+X(I,J)
         52 CONTINUE
            IF ( SUMJ .LE. 0. ) GO TO 60
            IF ( J .EQ. 7 ) GO TO 58
            IF ( SUMM-SUMJ ) 54,56,60
         54 M=J
            GO TO 58
         56 IF ( X(2,M) .LT. X(2,J) ) GO TO 60
            M=J
         58 SUMM=SUMJ
         60 SUMJ=0.
         62 CONTINUE
         64 N=3
            NN=NI-1
            DO 68 I=3,NN
            IF ( X(N,M) .LE. S ) GO TO 66
            R1=X(N,6)/X(N,M)
            IF ( X(I+1,M) .LE. S ) GO TO 68
            R2=X(I+1,6)/X(I+1,M)
            IF ( R1 .LE. R2 ) GO TO 68
```

```
FORTRAN IV G LEVEL  21            LP            DATE = 73097        11/11/21          PAGE C002

 66   N=I+1
 68   CONTINUE
      DO 70 I=1,4
 70   X(N,I)=X(NI+I,M)
      SAVE=X(N,M)
      PIVCT=1./SAVE
      X(N,M)=-1.
      DO 78 I=2,NI
      IF ( X(I,M)+S )  76,76,72
 72   IF  ( X(I,M)-S )  74,76,76
 74   W(I)=0.
      GO TO 78
 76   W(I)=-X(I,M)*PIVOT
 78   CONTINUE
      X(N,M)=SAVE
      DO 80 J=6,NJ
      TEMF=X(N,J)
      X(N,J)=0.
      DO 80 I=2,NI
      X(I,J)=X(I,J)+W(I)*TEMP
      IF ( J .EQ. M )  X(N,M)=1.
 80   CONTINUE
      DO 82 I=2,NI
      IF ( I .EQ. N )  GO TO 82
      X(I,M)=0.
 82   CONTINUE
      NIT=NIT+1
      NCT=0
      TES1=0.
 86   DC 88 I=3,NI
      IF ( B(I,2) .NE. X(I,2) )  GO TO 88
      IF ( B(I,3) .NE. X(I,3) )  GO TO 88
      IF ( B(I,4) .NE. X(I,4) )  GO TO 88
      TEST=TEST+X(I,6)
      NCT=NCT+1
 88   CONTINUE
      IF ( NIT .EQ. 1 )  GO TO 96
      IF ( NM .NE. M )  GO TO 96
      IF ( TEST1 .GT. TEST )  GC TO 96
 89   NI=NI-2
      WRITE (6,1)  NI,NIT
      NEBE=0
      GO TO 999
 96   TEST1=TEST
      NM=M
      IF ( NIT .GT. 100 )  GO TC 89
      IF ( NCT .NE. 0 )  GO TO 51
 98   M=7
      DO 106 J=8,NJ
      IF ( X(2,M)+S )  100,104,104
100   IF ( X(2,J)+S )  102,106,106
102   IF ( X(2,M) .LT. X(2,J) )  GO TO 106
104   M=J
106   CCNTINUE
```

```
FORTRAN IV G LEVEL  21                LP               DATE = 73097          11/11/21             PAGE 0003

               IF ( (X(2,M)+S) .LT. 0. ) GO TO 64
          999  RETURN
          1    FORMAT (//16H NO SOLUTION FOR,I4,7H MOMENT//)
               END
```

```
FORTRAN IV G LEVEL  21                 LP                 DATE = 73097          11/11/21            PAGE 0004

  *OPTIONS IN EFFECT*  ID,EBCDIC,SOURCE,NOLIST,NODECK,LOAD,NOMAP
  *OPTIONS IN EFFECT*  NAME = LP       , LINECNT =       56
  *STATISTICS*    SOURCE STATEMENTS =       105,PROGRAM SIZE =     2800
  *STATISTICS*  NO DIAGNOSTICS GENERATED
```

```
FORTRAN IV G LEVEL  21              MOMVAL          DATE = 73097          11/11/21             PAGE 0001

0001               SUBROUTINE MOMVAL (VAL,MCM,VALC)
             C
             C  CALCULATION OF BOUNDARIES FOR ESTIMATED CENTRAL MOMENTS:
             C  CONVERSION OF FREQUENCIES IN X MATRIX FROM ALPHA TO NUMERIC AND
             C  CALCULATION OF MOMENT
             C
             C  SUBROUTINE WRITTEN ORIGINALLY BY DR. RICHARD K. HILL, JR
             C  (VIRGINIA POLYTECHNIC INSTITUTE) AND MODIFIED BY DAVID M. SHOEMAKER
             C
0002               COMMON /BLOCK6/ SSUM(4),X(10,157),B(6,4),W(6)
0003               DIMENSION G(10)
0004               DATA G/1H0,1H1,1H2,1H3,1H4,1H5,1H6,1H7,1H8,1H9/
0005               NI=X(1,5)
0006               VAL=0.
0007               DO 97 I=3,NI
0008               DO 98 L=1,10
0009               IF ( X(I,2) .EQ. G(L) ) GO TO 99
0010          98   CONTINUE
0011          99   SAVE=(L-1)*100
0012               DO 100 L=1,10
0013               IF ( X(I,3) .EQ. G(L) ) GO TO 101
0014         100   CONTINUE
0015         101   SAVE=SAVE+(L-1)*10
0016               DO 102 L=1,10
0017               IF ( X(I,4) .EQ. G(L) ) GO TO 103
0018         102   CONTINUE
0019         103   SAVE=SAVE+L-1
0020          97   VAL=VAL+X(I,6)*(SAVE**MCM)
0021               V1=SSUM(1)
0022               V2=SSUM(2)
0023               V3=SSUM(3)
0024               K=MCM-1
0025               GO TO (200,300,400), K
0026         200   VALC=VAL-V1*V1
0027               RETURN
0028         300   VALC=VAL-3.*V1*V2+2.*V1**3
0029               RETURN
0030         400   VALC=VAL-4.*V3*V1+6.*V2*V1*V1-3.*V1**4
0031               RETURN
0032               END
```

```
FORTRAN IV G LEVEL  21              MOMVAL          DATE = 73097         11/11/21            PAGE 0002

  *OPTIONS IN EFFECT*  ID,EBCDIC,SOURCE,NOLIST,NODECK,LOAD,NOMAP
  *OPTIONS IN EFFECT*  NAME = MOMVAL  , LINECNT =       56
  *STATISTICS*    SOURCE STATEMENTS =       32,PROGRAM SIZE =     1136
  *STATISTICS*  NO DIAGNOSTICS GENERATED
```

```
FORTRAN IV G LEVEL  21              NEGHGR            DATE = 73097          11/11/21               PAGE 0001

0001                  SUBFOUTINE NEGHGR (KPOP,XEAR,XVAR)
               C
               C   GRADUATION OF NORMATIVE DISTRIBUTION BY NEGATIVE HYPERGEOMETRIC
               C   DISTRIEUTICN
               C
               C   COMPUTATION OF CONSTANTS
               C
0002                  A21=(KPOP/(KPOP-1.))*(1.-XEAR*(KPOP-XBAR)/(KPOP*XVAR))
0003                  WRITE (6,1) XBAR,XVAR,A21
0004                  IF ( A21 .GT. 0. .AND. A21 .LT. 1. ) GO TO 120
0005                  WRITE (6,2)
0006                  RETURN
0007             120  A=(-1.+1./A21)*XEAR
0008                  B=-A-1.+KPOP/A21
0009                  SLCG1=0.
0010                  SLOG2=0.
0011                  C=A+B
0012                  DO 140 I=1,KPOP
0013                  SLCG1=SLOG1+ALOG10(B-I+1.)
0014             140  SLOG2=SLOG2+ALOG10(C-I+1.)
0015                  C=10.**(SLOG1-SLOG2)
               C
               C   GENERATION OF NEGATIVE HYPERGEOMETRIC DISTRIBUTION
               C
0016                  WRITE (6,3)
0017                  S=KEOP
0018                  N3=KPOP+1
0019                  CK=0.
0020                  DO 200 J=1,N3
0021                  K=J-1
0022                  IF ( K .EQ. 0 ) GO TO 160
0023                  SLOG1=0.
0024                  SLCG2=0.
0025                  SLOG3=0.
0026                  SLOG4=0.
0027                  DO 150 I=1,K
0028                  SLOG1=SLOG1+ALOG10(S-I+1.)
0029                  SLOG2=SLOG2+ALOG10(A+I-1.)
0030                  SLCG3=SLOG3+ALOG10(B-I+1.)
0031             150  SLOG4=SLOG4+ALOG10(FLOAT(I))
0032                  HX=C*10.**(SLOG1+SLOG2-SLCG3-SLOG4)
0033                  GO TO 165
0034             160  HX=C
0035             165  CK=CK+HX
0036                  WRITE (6,4) K,HX,CK
0037             200  CONTINUE
               C
0038                  RETURN
               C
0039               1  FORMAT (//82H APPROXIMATICN OF NORMATIVE DISTRIBUTION WITH NEGATIV
                     1E HYPERGEOMETRIC DISTRIBUTION//21H ESTIMATED PARAMETERS//5H MEAN,
                     2F25.5/9H VARIANCEF21.5/4H A21F26.5)
0040               2  FORMAT (//30H DISTRIBUTICN ABORTED A21 LE 0//)
0041               3  FORMAT (//5X,5HINDEX,5X,13HRELATIVE FREQ,10X,12HCUM REL FREQ//)
```

```
FORTRAN IV G LEVEL  21              NEGHGR          DATE = 73097          11/11/21            PAGE 0002

 0042          4     FORMAT (1X,I9,F18.5,F22.5)
 0043                END
```

```
FORTRAN IV G LEVEL  21              NEGHGR          DATE = 73097          11/11/21            PAGE 0003

 *OPTIONS IN EFFECT*  ID,EBCDIC,SOURCE,NOLIST,NODECK,LOAD,NOMAP
 *OPTIONS IN EFFECT*  NAME = NEGHGR  , LINECNT =       56
 *STATISTICS*     SOURCE STATEMENTS =        43,PROGRAM SIZE =      1848
 *STATISTICS*  NO DIAGNOSTICS GENERATED
```

```
FORTRAN IV G LEVEL  21              PEARSN            DATE = 73097          11/11/21              PAGE 0001

0001              SUBROUTINE PEARSN(U1,U2,U3,U4,L1,L2,STEP)
            C
            C
            C  GRADUATION OF FREQUENCY DISTRIBUTIONS WITH PEARSON CURVES
            C  WITH RELATIVE FREQUENCIES PEE INDEX VALUE CALCULATED
            C
            C
            C  ARGUMENTS OF SUEFOUTINE
            C
            C     U1    MEAN
            C     U2    2ND CENTRAL MOMENT (VARIANCE)
            C     U3    3RD CENTRAL MCMENT
            C     U4    4TH CENTRAL MOMENT
            C     L1    LOWER LIMIT CF INDEX
            C     L2    UPPER LIMIT OF INDEX
            C   STEP    DESIRED STEP SIZE IN GOING FRCM L1 TO L2
            C
            C  REQUIRED SUBROUTINES
            C
            C     GAMMA   EXP   ABS   SQRT   SIN   COS   ATAN
            C
0002              REAL M,M1,M2,L1,L2,LAMBDA
0003              DIMENSION CF(4),RR(4),RI(4)
            C
            C  COMPUTATION OF INDEX OF SKEWNESS (B1)  AND DEGREE OF KURTOSIS (B2)
            C  AND CRITERION VALUE USED IN THE SELECTION OF APPROPRIATE PEARSON
            C  CURVE
            C
0004              X0=0.
0005              XN=C.
0006              NHIGH=0
0007              CUMF=0.
0008              PI=3.1415927
0009              TPI=2.*PI
0010              B1=U3*U3/(U2**3)
0011              B2=U4/(U2*U2)
0012              Z1=B1
0013              Z2=B2
0014              CRIT1=(B1*(B2+3.)**2)/(4.*(4.*B2-3.*B1)*(2.*B2-3.*B1-6.))
0015              WRITE (6,14) U1,U2,U3,U4
0016              WRITE (6,1) B1,B2,CRIT1
0017              IF ( U2 .GT. 0. ) GO TO 60
0018              WRITE (6,15)
0019              RETURN
0020         60   IF ( L1 .LT. L2 .AND. STEP .GT. 0. ) GO TO 65
0021              WRITE (6,16)
0022              RETURN
0023         65   CONTINUE
0024              A=4.*B2-3.*B1
0025              B=3.*B1-2.*B2+6.
0026              CRIT2=(A*(10.*B2-12.*B1-18.)**2-B1*(B2+3.)**2*(8.*B2-9.*B1-12.))/
                 1       (B*(B1*(B2+3.)**2+4.*A*B))
0027              CRIT3=5.*B2-6.*B1-9.
0028              CRIT4=2.*B2-3.*B1-6.
```

```
FORTRAN IV G LEVEL  21                PEARSN           DATE = 73097        11/11/21              PAGE 0002

0029               WRITE (6,23) CRIT2,CRIT3,CRIT4
             C
             C  TEST FOR VALUES OF CRITERIA APPROXIMATING THEORETICAL VALUES
             C
0030               IF ( CRIT1.LE. .02 .AND. CRIT1.GE. -.02 ) CRIT1=0.
0031               IF ( CRIT1.LE. 1.02 .AND. CRIT1.GE. .98 ) CRIT1=1.
0032               IF ( CRIT2 .GE. -.02 .AND. CRIT2 .LE. .02 ) CRIT2=0.
0033               IF ( CRIT3 .GE. -.02 .AND. CRIT3 .LE. .02 ) CRIT3=0.
0034               IF ( CRIT4 .GT. -.02 .AND. CRIT4 .LT. .02 ) CRIT4=0.
0035               IF ( B1 .GE. 3.98 .AND. B1 .LE. 4.02 ) Z1=4.
0036               IF ( B2 .LE. 3.02 .AND. B2 .GE. 2.98 ) Z2=3.
0037               IF ( B2 .GE. 8.98 .AND. B2 .LE. 9.02 ) Z2=9.
             C
             C  TEST FOR CURVE BEING MAIN TYPE OR TRANSITION TYPE
             C
0038               IF ( CRIT4 .EQ. 0. ) GO TO 600
0039               IF ( Z1 .EQ. 4. .AND. Z2 .EQ. 9. ) GO TO 800
0040               IF ( CRIT2 ) 80,79,80
0041          79   IF ( CRIT1 .LT. 0. .AND. CRIT3 .LT. 0. ) GO TO 700
0042               IF ( CRIT1 .GT. 1. .AND. CRIT4 .GT. 0. ) GO TO 850
0043               IF ( CRIT1.LT.0..AND.CRIT3.GT.0..AND.CRIT4.LT.0. ) GO TO 750
0044          80   IF ( CRIT1) 100,400,90
0045          90   IF ( CRIT1-1.) 200,500,300
             C
             C  CURVE I *FIRST MAIN TYPE*
             C
0046         100   WRITE (6,4)
0047               R=6.*(B2-B1-1.)/(6.+3.*B1-2.*B2)
0048               C=B1*(R+2.)**2+16.*(R+1.)
0049               A12=SQRT(U2*C)/2.
0050               C=R*(R+2.)*SQRT(B1/C)
0051               M1=(R-2.-C)/2.
0052               M2=(R-2.+C)/2.
             C
             C  IF U3 IS POSITIVE, M2 IS POSITIVE ROOT
             C
0053               IF ( U3 .LE. 0. ) GO TO 120
0054               IF ( M2 .GE. M1 ) GO TO 120
0055               TEMP=M2
0056               M2=M1
0057               M1=TEMP
0058         120   A1=(M1+1.)*A12/(M1+M2+2.)
0059               A2=A12-A1
0060               YE=(1./A12)*(((M1+1.)**M1*(M2+1.)**M2)/(M1+M2+2.)**(M1+M2))*
                  1(GAMMA(M1+M2+2.)/(GAMMA(M1+1.)*GAMMA(M2+1.)))
0061               WRITE (6,2)
0062               X=L1
0063               KK=1
0064         150   IF ( X .GT. L2 ) GO TO 1000
0065               T=X-U1
0066               R1=1.+T/A1
0067               R2=1.-T/A2
0068               IF ( R1 .LE. 0. .AND. R2 .GT. 0. ) P=YE*R2**M2
0069               IF ( R2 .LE. 0. .AND. R1 .GT. 0. ) P=YE*R1**M1
```

```
FORTRAN IV G LEVEL  21              PEARSN            DATE = 73097         11/11/21              PAGE C003

 0070             IF ( R1 .LE. 0. .AND. R2 .LE. 0. ) P=YE
 0C71             IF ( R1 .GT. 0. .AND. R2 .GT. 0. ) P=YE*R1**M1*R2**M2
 0C72             IF ( R1 .LE. 0. .OR. R2 .LE. 0. ) NHIGH=1
 0C73             GO TC 5000
            C
            C  CURVE IV *SECOND MAIN TYPE*
            C
 0C74        200  WRITE (6,5)
 0075             R=6.*(B2-B1-1.)/(2.*B2-3.*B1-6.)
 0076             M=(R+2.)/2.
 0077             A=SQRT((U2/16.)*(16.*(R-1.)-B1*(R-2.)**2))
 0078             V=SQRT((R*R*(R-2.)**2*B1)/(16.*(R-1.)-B1*(R-2.)**2))
            C
            C  V AND A HAVE OPPOSITE SIGN OF U3
            C
 0079             SGN=1.
 0080             IF ( U3 .GT. 0. ) SGN=-1.
 0081             A=SGN*ABS(A)
 0C82             V=SGN*ABS(V)
            C
            C  NOTE ... ARCTAN(-Z)=-ARCTAN(Z)
            C
 0C83             THETA=ATAN(ABS(V/R))
 0C84             IF ( V/R .LT. 0. ) THETA=-THETA
 0C85             COST=CCS(AES(THETA))
            C
            C
 0086             IF ( V-2. ) 225,225,250
            C
            C
 0C87        225  YO=1./(AES(A)*FRV(R,V))
 0088             GO TC 270
            C
            C  APPROXIMATION FOR YO
            C
 0089        250  TEMP=(COST      **2)/(3.*R)-1./(12.*R)-THETA*V
 0090             YC=(1./A)*SQRT(R/TPI)*EXP(TEMP)/(COST       **(R+1.))
            C
            C
 0091        270  CONTINUE
 0C92             WRITE (6,2)
 0093             X=L1
 CC94             Y=V/R
 0C95             KK=2
 C096        280  IF ( X .GT. L2 ) GO TO 1000
 0097             T=X-U1
 0098             Q=T/A-Y
 0099             P=YO*(1.+Q*Q)**(-M)*EXP(-V*ATAN(Q))
 0100             GC TC 5000
            C
            C  CURVE VI *THIRD MAIN TYPE*
            C
 0101        300  WRITE (6,6)
 0102             R=6.*(B2-B1-1.)/(6.+3.*B1-2.*B2)
```

```
FORTRAN IV G LEVEL  21                PEARSN            DATE = 73097          11/11/21            PAGE C004

 0103              C=B1*(R+2.)**2+16.*(R+1.)
 0104              A=SQRT(U2*C)/2.
             C
             C  A HAS SAME SIGN AS U3
             C
 0105              IF ( U3 .LT. 0. ) A=-A
 0106              TEMP=R*(R+2.)/2.*SQRT(B1/C)
 0107              Q1=-(R-2.)/2.+TEMP
 0108              Q2=(R-2.)/2.+TEMP
             C
             C  R IS ALWAYS NEGATIVE HERE.  Q1 IS GREATER THAN Q2
             C
 0109              T1=Q2+1.
 0110              T2=Q1-Q2-2.
 0111              T3=Q1-1.
 0112              A1=A*T3/T2
 0113              A2=A*T1/T2
 0114              YE=(T1**Q2*T2**(Q1-Q2)*GAMMA(Q1))/(ABS(A)
                  1  *T3**Q1*GAMMA(T2+1.)*GAMMA(T1))
 0115              WRITE (6,2)
 0116              X=L1
 0117              KK=3
 0118         350  IF ( X .GT. L2 ) GO TO 1000
 0119              T=X-U1
 0120              R1=ABS(1.+T/A1)
 0121              R2=ABS(1.+T/A2)
 0122              P=YE*R1**(-Q1)*R2**Q2
 0123              GO TO 5000
             C
             C  TRANSITION CURVES
             C
 0124         400  IF ( Z2 -3. ) 450,410,480
             C
             C  NORMAL CURVE OF ERROR *TRANSITION CURVE*
             C
 0125         410  WRITE (6,7)
 0126              YO=1./SQRT(TPI*U2)
 0127              WRITE (6,2)
 0128              X=L1
 0129              KK=4
 0130         420  IF ( X .GT. L2 ) GO TO 1000
 0131              P=YO*EXP(-((X-U1)**2)/(2.*U2))
 0132              GO TO 5000
             C
             C  TYPE II CURVE *TRANSITION CURVE*
             C
 0133         450  WRITE (6,8)
 0134              M=(5.*B2-9.)/(6.-2.*B2)
 0135              A2=(2.*U2*B2)/(3.-B2)
 0136              YO=(1./SQRT(A2*PI))*GAMMA(M+1.5)/GAMMA(M+1.)
 0137              WRITE (6,2)
 0138              X=L1
 0139              KK=5
 0140         460  IF ( X .GT. L2 ) GO TO 1000
```

```
FORTRAN IV G LEVEL  21              PEARSN           DATE = 73097          11/11/21            PAGE 0005

0141             T=X-U1
0142             R1=1.-T*T/A2
0143             IF ( R1 .LE. 0. ) P=YO
0144             IF ( R1 .GT. 0. ) P=YO*R1**M
0145             IF ( R1 .LE. 0. ) NHIGH=1
0146             GO TC 5000
          C
          C  TYPE VII CURVE *TRANSITION CURVE*
          C
0147        480  WRITE (6,9)
0148             M=(5.*B2-9.)/(2.*B2-6.)
0149             A2=(2.*U2*B2)/(B2-3.)
0150             YO=(1./SQRT(A2*PI))*GAMMA(M)/GAMMA(M- .5)
0151             WRITE (6,2)
0152             X=L1
0153             KK=6
0154        490  IF ( X .GT. L2 ) GO TO 1000
0155             T=X-U1
0156             P=YO*(1.+T*T/A2)**(-M)
0157             GO TC 5000
          C
          C  TYPE V CURVE *TRANSITION CURVE*
          C
0158        500  WRITE (6,10)
0159             P=4.+(8.+4.*SQRT(4.+B1))/B1
0160             G=(P-2.)*SQRT(U2*(P-3.))
0161             YC=G**(P-1.)/GAMMA(P-1.)
0162             ORIGIN=U1-G/(P-2.)
0163             IF ( U3 .LT. 0. ) ORIGIN=U1+G/(P-2.)
0164             WRITE (6,2)
0165             X=L1
0166             KK=7
0167        520  IF ( X .GT. L2 ) GO TO 1000
0168             X=X-CRIGIN
0169             T=ABS(X)
0170             P=YO*T**(-P)*EXP(-G/X)
0171             GO TO 5000
          C
          C  TYPE III CURVE *TRANSITION CUBVE*
          C
0172        600  IF ( B1 .EQ. 0. ) GO TO 80
0173             WRITE (6,11)
0174             LAMBDA=2.*U2/U3
0175             P=4./B1-1.
0176             A=(P+1.)/LAMBDA
0177             YE=LAMBDA*(((P+1.)**P)/(EXP(P+1.)*GAMMA(P+1.)))
0178             WRITE (6,2)
0179             X=L1
0180             KK=8
0181        620  IF ( X .GT. L2 ) GO TO 1000
0182             T=X-U1
0183             R1=1.+T/A
0184             IF ( R1 .LE. 0. ) P=YE*EXP(-LAMBDA*T)
0185             IF ( R1 .GT. 0. ) P=YE*R1**P*EXP(-LAMBDA*T)
```

```
FORTRAN IV G LEVEL  21                PEARSN          DATE = 73097          11/11/21             PAGE 0006

0186               IF ( R1 .LE. 0. ) NHIGH=1
0187               GO TO 5000
             C
             C  TYPE VIII CURVE *TRANSITION CURVE*
             C
C188           700  WRITE (6,19)
             C
             C  SOLUTICN OF CUBIC EQUATION FOE VALUE OF M
             C
             C  WITH TYPE VIII, M IS RESTRICTED TO 0-1 RANGE
             C
0189               CF(1)=16.*B1
0190               CF(2)=-24.*B1
0191               CF(3)=9.*B1-12.
0192               CF(4)=4.-B1
0193               CALL POLRT (CF,3,RR,RI,IER)
0194               DO 7C5 I=1,3
0195               M=RR(I)
0196               IF (M.GE.0..AND.M.LE.1..ANC.RI(I).EQ.0. ) GO TO 710
0197           705  CONTINUE
0198               WRITE (6,25)
0199               GO TO 750
0200           710  T=(2.-M)*SQRT(U2*(3.-M)/(1.-M))
0201               A=T*(1.-M)/(2.-M)
             C
             C  A HAS CPPOSITE SIGN OF U3
             C
0202               IF ( U3 .LT. 0. ) A=-A
C203               YE=((1.-M)*(2.-M)**M)/(T*(1.-M)**M)
0204               WRITE (6,2)
0205               X=L1
02C6               KK=9
0207           720  IF ( X .GT. L2 ) GO TO 1000
0208               T=X-U1
0209               P=YE*(1.-T/A)**(-M)
0210               GO TC 5000
             C
             C  TYPE IX CURVE *TRANSITION CURVE*
             C
0211           750  WRITE (6,20)
             C
             C  SOLUTICN OF CUBIC EQUATION FOE VALUE OF M
             C
             C  WITH TYPE IX, M MUST BE NON-NEGATIVE
             C
0212               CF(1)=16.*B1
0213               CF(2)=24.*B1
0214               CF(3)=9.*B1-12.
0215               CF(4)=B1-4.
0216               CALL PCLRT (CF,3,RR,RI,IEF)
0217               DO 755 I=1,3
0218               M=RR(I)
0219               IF ( M .GE. 0. .AND. RI(I) .EQ. 0. ) GO TO 769
0220           755  CONTINUE
```

```
FORTRAN IV G LEVEL  21          PEARSN          DATE = 73097          11/11/21          PAGE C007

0221              WRITE (6,26)
0222              GO TO 100
0223          769 T=(M+2.)*SQRT(U2*(M+3.)/(M+1.))
0224              A=(M+1.)*T/(M+2.)
0225              YE=(M+1.)**(M+1.)/(T*(M+2.)**M)
0226              WRITE (6,2)
0227              X=L1
0228              KK=10
0229          770 IF ( X .GT. L2 ) GO TO 1000
0230              T=X-U1
0231              P=YE*(1.+T/A)**M
0232              GO TO 5000
            C
            C  TYPE X CURVE *TRANSITION CURVE*
            C
0233          800 WRITE (6,21)
0234              A=SQRT(U2)
0235              YE=1./(2.718282*A)
0236              WRITE (6,2)
0237              X=L1
0238              KK=11
0239          820 IF ( X .GT. L2 ) GO TO 1000
0240              T=(X-U1)
0241              IF ( U3 .LT. 0. ) P=YE*EXE(T/A)
0242              IF ( U3 .GE. 0. ) P=YE*EXP(-T/A)
0243              GO TO 5000
            C
            C  TYPE XI CURVE *TRANSITION CURVE*
            C
0244          850 WRITE (6,22)
            C
            C  SOLUTICN OF CUBIC EQUATION FCB VALUE CF M
            C
            C  WITH TYPE XI, M MUST BE GREATER THAN 1
            C
0245              CF(1)=16.*B1
0246              CF(2)=-24.*B1
0247              CF(3)=9.*B1-12.
0248              CF(4)=4.-B1
0249              CALL POLRT (CF,3,RR,RI,IER)
0250              DO 855 I=1,3
0251              M=RR(I)
0252              IF ( M .GT. 1. .AND. RI(I) .EQ. 0. ) GO TO 869
0253          855 CONTINUE
0254              WRITE (6,24)
0255              GO TO 300
0256          869 B=(M-2.)*SQRT(U2*(M-3.)/(M-1.))
0257              YE=((M-2.)**M)/(B*(M-1.)**(M-1.))
            C
            C  B HAS SAME SIGN AS U3
            C
0258              B=ABS(B)
0259              IF ( U3 .LT. 0. ) B=-B
0260              A=B*(M-1.)/(M-2.)
```

```
FORTRAN IV G LEVEL  21                PEARSN            DATE = 73097          11/11/21                PAGE 0008

          WRITE (6,2)
          X=L1
          KK=12
  870     IF ( X .GT. L2 ) GO TO 1000
          T=X-U1
          P=YE*(1.+T/A)**(-M)
          GO TO 5000
C
C  AREA UNDER CURVE APPROXIMATED BY TRAPEZOID FORMULA
C
 1000 AREA=STEP*(CUMP-(X0+XN)/2.)
          WRITE (6,13) AREA
          RETURN
C
C
 5000 WRITE (6,3) X,P
          IF ( P .GE. 1. ) P=0.
          CUMP=CUMP+P
          IF ( NHIGH .EQ. 0 ) GO TO 5010
          WRITE (6,18)
          NHIGH=0
 5010 IF ( X .EQ. L1 ) X0=P
          IF ( X .EQ. L2 ) XN=P
          X=X+STEP
          GO TO (150,280,350,420,460,490,520,620,720,770,820,870), KK
C
C
    1     FORMAT (   /18H INDEX OF SKEWNESS F21.5/19H DEGREE OF KURTOSIS
         1F20.5/16H CRITERION VALUEF23.5)
    2     FORMAT (5X,5HINDEX,5X,13HRELATIVE FREQ//)
    3     FORMAT (1X,F9.3,F18.5)
    4     FORMAT (/22H TYPE I CURVE SELECTED/)
    5     FORMAT (/23H TYPE IV CURVE SELECTED/)
    6     FORMAT (/23H TYPE VI CURVE SELECTED/)
    7     FORMAT (/28H NORMAL ERROR CURVE SELECTED/)
    8     FORMAT (/23H TYPE II CURVE SELECTED/)
    9     FORMAT (/24H TYPE VII CURVE SELECTED/)
   10     FORMAT (/22H TYPE V CURVE SELECTED/)
   11     FORMAT (/24H TYPE III CURVE SELECTED/)
   13     FORMAT (//15H APPROX AREA ISF15.7//)
   14     FORMAT (////49H APPROXIMATION OF DISTRIBUTION WITH PEARSON CURVE//
         121H ESTIMATED PARAMETERS///5H MEANF34.5/9H VARIANCEF30.5/19H 3RD C
         2ENTRAL MOMENT F20.5/19H 4TH CENTRAL MOMENTF20.5)
   15     FORMAT (//23H NO CURVE WHEN U(2) = 0//)
   16     FORMAT (//38H INVALID LIMITS AND/OR DELTA ARGUMENTS//)
   18     FORMAT (1H+,30X,32HORDINATE UNDEFINED AT THIS POINT)
   19     FORMAT (/25H TYPE VIII CURVE SELECTED/)
   20     FORMAT (/23H TYPE IX CURVE SELECTED/)
   21     FORMAT (/22H TYPE X CURVE SELECTED/)
   22     FORMAT (/23H TYPE XI CURVE SELECTED/)
   23     FORMAT (/7H LAMBDA,E36.5/10H 5B2-6B1-9,F29.5/10H 2B2-3B1-6,F29.5/)
   24     FORMAT (/22H M .LE. 1. FOR TYPE XI /)
   25     FORMAT (/29H M OUT OF RANGE FOR TYPE VIII /)
   26     FORMAT (/23H M NEGATIVE FOR TYPE IX /)
```

```
FORTRAN IV G LEVEL  21              PEARSN          DATE = 73097        11/11/21           PAGE 0009
 0305              END
```

```
FORTRAN IV G LEVEL  21                PEARSN          DATE = 73097         11/11/21           PAGE 0010

  *OPTIONS IN EFFECT*  ID,EBCDIC,SOURCE,NOLIST,NODECK,LOAD,NOMAP
  *OPTIONS IN EFFECT*  NAME = PEARSN  , LINECNT =       56
  *STATISTICS*    SOURCE STATEMENTS =       305,PROGRAM SIZE =    10496
  *STATISTICS*  NO DIAGNOSTICS GENERATED
```

```
FORTRAN IV G LEVEL  21              POOL            DATE = 73097          11/11/21              PAGE 0001

0001              SUBROUTINE POOL (NSPT,IPT,NT,SX,SXX,SEU1,SEU2)
             C
             C  CALCULATION OF POOLED ESTIMATES OF PARAMETERS FROM SUBTEST DATA
             C
             C
             C  NOTE:
             C
             C  150 = MAXIMUM NUMBER OF ITEMS PER SUBTEST
             C  100 = MAXIMUM NUMBER OF EXAMINEES PER SUBTEST
             C
0002              INTEGER FQ
0003              REAL MPOP
0004              COMMON P(150),X(150),LT(1500)
0005              COMMON /BLOCK1/ YBAR,YSD,MPOP,SPOP,KPOP,NDIST,IPRT
0006              COMMON /BLOCK7/ FQ(52,13),CFQ(5,13),NFREQ
0007              DIMENSION TEST(150),XX(100,150),SX(13),SXX(13),Z(8),CONF(6),
                 1    PRCNT(6)
0008              DATA CONF/4H70.0,4H80.0,4H90.0,4H95.0,4H97.5,4H99.0/
0009              DATA PRCNT/.70,.80,.90,.95,.975,.99/
0010              IF ( IPRT .EQ. 1 ) WRITE (6,1)
             C
             C
0011              IF ( IPT .LE. 150 .AND. NSPT .LE. 100 ) GO TO 50
0012              WRITE (6,7)
0013              IF ( IPT .GT. 150 ) IPT=150
0014              IF ( NSPT .GT. 100 ) NSPT=100
0015          50  CONTINUE
0016              VAR1=0.
0017              VAR2=0.
0018              SVCIT=0.
0019              DO 100 I=1,13
0020              SX(I)=0.
0021         100  SXX(I)=0.
             C
             C
0022              DO 1000 I=1,NT
             C
             C
0023              ISTART=IPT*(I-1)+1
0024              ISTOP=IPT*I
0025              DO 500 L=1,NSPT
0026              CALL DATA (TEST)
0027              LL=0
0028              DO 500 K=ISTART,ISTOP
0029              KK=LT(K)
0030              LL=LL+1
0031         500  XX(L,LL)=TEST(KK)
             C
0032              CALL ESTORS (XX,NSPT,IPT,2,VU1,VU2)
0033              CALL VCOV (XX,NSPT,IPT,VCOVIT)
             C
0034              VAR1=VAR1+VU1
0035              VAR2=VAR2+VU2
0036              SVCIT=SVCIT+VCOVIT
```

```
FORTRAN IV G LEVEL  21                POOL              DATE = 73097          11/11/21              PAGE 0002

0037                DO 700 J=1,8
0038                SX(J)=SX(J)+Z(J)
0039            700 SXX(J)=SXX(J)+Z(J)*Z(J)
               C
               C
0040                IF ( IPRT .EQ. 1 ) WRITE (6,2) I, (Z(J),J=1,8)
               C
               C
0041           1000 CONTINUE
               C
               C  COMPUTE JACKKNIFE STANDARD ERRORS OF ESTIMATE
               C
0042                NN=NSPT*NT
0043                IF ( NT .EQ. 1 ) GO TO 1111
0044                DO 1110 I=1,8
0045                TEMP=((SXX(I)-SX(I)**2/NT)/(NT-1.))/NT
0046                IF ( TEMP .GT. 0. ) GO TO 1109
0047                SXX(I)=0.
0048                GO TO 1110
0049           1109 SXX(I)=SQRT(TEMP)
0050           1110 SX(I)=SX(I)/NT
               C
               C  COMPUTATION OF SE FOR POOLED ESTIMATE OF MEAN AND VARIANCE
               C  USING THE HOOKE-LORD-NOVICK PROCEDURE
               C
0051           1111 SAVE=VAR1
0052                VAR1=VAR1/(NT*NT)-(KPOP*KECP*(NT-1.)*SX(5))/(NT*(KPOP-1.))
0053                IF ( NT*IPT .GT. KPOP ) VAR1=SAVE/NT**2
0054                VAR2=VAR2/(NT*NT)-(4.*(NT-1.)*SVCIT*KPOP)/(NT*NT)
0055                SEU1=0.
0056                SEU2=0.
0057                IF ( VAR1 .GT. 0. ) SEU1=SQRT(VAR1)
0058                IF ( VAR2 .GT. 0. ) SEU2=SQRT(VAR2)
               C
0059                IF ( IPRT .EQ. 1 ) WRITE (6,3) (SX(I),I=1,8),(SXX(I),I=1,8)
               C
0060                IF ( NFREQ .EQ. 1 .AND. IPRT .EQ. 0 ) WRITE (6,3)
                   1                                  (SX(I),I=1,8),(SXX(I),I=1,8)
               C
0061                IF ( IPRT .EQ. 1 ) WRITE (6,8) VAR1,VAR2
               C
               C
0062                IF ( NT .EQ. 1 ) RETURN
               C
               C  COMPUTE CONFIDENCE INTERVALS FOR PARAMETERS
               C  (DF = NUMBER OF SUBTESTS MINUS ONE)
               C
0063                IF ( IPRT .EQ. 0 ) RETURN
0064                WRITE (6,6)
0065                DO 2000 J=3,6
0066                CALL TTABLE (T,NT-1,PRCNT(J))
0067                WRITE (6,4) CONF(J)
0068                DO 1400 I=1,8
0069                TEMP=T*SXX(I)
```

```
FORTRAN IV G LEVEL  21                  POOL              DATE = 73097          11/11/21               PAGE 0003

0070              Z(I)=SX(I)-TEMP
0071         1400 P(I)=SX(I)+TEMP
0072              WRITE (6,5)  (Z(I),I=1,8),(P(I),I=1,8)
0073         2000 CONTINUE
            C
            C
0074              RETURN
            C
            C
0075          1   FORMAT (///25X,21HESTIMATES OF MOMENTS ,33X,
                 122HCOMPONENTS OF VARIANCE/8X,
                 257H---------------------------------------------------------,3X
                 3,57H---------------------------------------------------------/
                 48H SUBTEST,5X,4HMEAN, 9X,8HVARIANCE,10X,2HU3,13X,2HU4,
                 511X,6HVAR(I), 9X,6HVAR(E), 8X,7HVAR(IE), 9X,6HTHETA'//)
0076          2   FORMAT (1X,I4,8E15.5)
0077          3   FORMAT (//5H MEAN8E15.5/5H SE  8E15.5)
0078          4   FORMAT (1X,A4,1X,27HPERCENT CONFIDENCE INTERVAL)
0079          5   FORMAT (5H FROM8E15.5/5H TO  8E15.5//)
0080          6   FORMAT (///)
0081          7   FORMAT (//57H *** N OR K PER SUBTEST OUT OF RANGE: PARAMETERS MODI
                 1FIED//)
0082          8   FORMAT (//12H VAR(U1)-HLN,F15.5/12H VAR(U2)-HLN,F15.5)
0083              END
```

```
FORTRAN IV G LEVEL  21                POOL              DATE = 73097          11/11/21             PAGE 0004

  *OPTIONS IN EFFECT*  ID,EBCDIC,SOURCE,NOLIST,NODECK,LOAD,NOMAP
  *OPTIONS IN EFFECT*  NAME = POOL    , LINECNT =       56
  *STATISTICS*    SOURCE STATEMENTS =       83,PROGRAM SIZE =    64376
  *STATISTICS*  NO DIAGNOSTICS GENERATED
```

```
FORTRAN IV G LEVEL  21              POLRT              DATE = 73097          11/11/21                PAGE 0001

0001              SUBROUTINE POLRT (XCOF,M,ROOTR,ROOTI,IER)
            C
            C  COMPUTATION OF THE REAL AND COMPLEX ROOTS OF A REAL POLYNOMIAL
            C  METHOD:  NEWTON-RAPHSON ITERATIVE TECHNIQUE.  THE FINAL ITERATIONS ON
            C          EACH ROOT ARE PERFORMED USING THE ORIGINAL POLYNOMIAL RATHER
            C          THAN THE REDUCED POLYNOMIAL TO AVOID ACCUMULATED ERRORS IN THE
            C          POLYNOMIAL.
            C
            C  LIMITATIONS:  LIMITED TO 36TH ORDER POLYNOMIAL OR LESS
            C
            C                FLOATING POINT OVERFLOW MAY OCCUR FOR HIGH ORDER
            C                POLYNOMIALS BUT WILL NOT AFFECT THE ACCURACY OF THE
            C                RESULTS.
            C
            C  PARAMETERS
            C
            C  XCOF     VECTOR OF M+1 COEFFICIENTS OF THE POLYNOMIAL ORDERED FROM
            C              SMALLEST TO LARGEST POWER
            C  COF      WORKING VECTOR OF LENGTH M+1
            C  M        ORDER OF POLYNOMIAL
            C  ROOTR    RESULTANT VECTOR OF LENGTH M CONTAINING REAL ROOTS
            C  ROOTI    RESULTANT VECTOR OF LENGTH M CONTAINING THE CORRESPONDING
            C            IMAGINARY ROOTS
            C  IER      ERROR CODE RETURNED WHERE
            C              0 = NO ERROR
            C              1 = M LESS THAN 1
            C              2 = M GREATER THAN 36
            C              3 = UNABLE TO DETERMINE ROOT WITH 500 ITERATIONS ON 5
            C                  STARTING VALUES
            C              4 = HIGH ORDER COEFFICIENT IS ZERO
            C  SUBROUTINE TAKEN FROM IBM/SSP
0002              DIMENSION XCOF(1),COF(36),ROOTR(1),ROOTI(1)
0003              IFIT=0
0004              N=M
0005              IER=0
0006              IF ( XCOF(N+1) )  10,25,10
0007         10   IF  ( N )  15,15,32
            C
            C  SET ERROR CODE TO 1
            C
0008         15   IER=1
0009         20   RETURN
            C
            C  SET ERROR CODE TO 4
            C
0010         25   IER=4
0011              GO TO 20
            C
            C  SET ERROR CODE TO 2
            C
0012         30   IER=2
0013              GO TO 20
0014         32   IF ( N-36 ) 35,35,30
0015         35   NX=N
```

```
FORTRAN IV G LEVEL  21              POLRT              DATE = 73097          11/11/21            PAGE 0002

0016                 NXX=N+1
0017                 N2=1
0018                 KJ1=N+1
0019                 DO 40 L=1,KJ1
0020                 MT=KJ1-L+1
0021           40    COF(MT)=XCOF(L)
                C
                C  SET INITIAL VALUES
                C
0022           45    XO=.5001010E-2
0023                 YO=.1000101E-1
                C
                C  ZERO INITIAL VALUE COUNTER
                C
0024                 IN=0
0025           50    X=XC
                C
                C  INCREMENT INITIAL VALUES AND COUNTER
                C
0026                 XO=-10.*YO
0027                 YO=-10.*X
                C
                C  SET X AND Y TO CURRENT VALUE
                C
0028                 X=XO
0029                 Y=YC
0030                 IN=IN+1
0031                 GO TC 59
0032           55    IFIT=1
0033                 XPR=X
0034                 YPR=Y
                C
                C  EVALUATE POLYNOMIAL AND DERIVATIVES
                C
0035           59    ICT=0
0036           60    UX=0.
0037                 UY=0.
0038                 V=0.
0039                 YT=0.
0040                 XT=1.
0041                 U=COF(N+1)
0042                 IF ( U ) 65,130,65
0043           65    DO 70 I=1,N
0044                 L=N-I+1
0045                 XT2=X*XT-Y*YT
0046                 YT2=X*YT+Y*XT
0047                 U=U+COF(L)*XT2
0048                 V=V+COF(L)*YT2
0049                 FI=I
0050                 UX=UX+FI*XT*COF(L)
0051                 UY=UY-FI*YT*COF(L)
0052                 XT=XT2
0053           70    YT=YT2
0054                 SUMSQ=UX*UX+UY*UY
```

```
FORTRAN IV G LEVEL  21              POLRT           DATE = 73097        11/11/21          PAGE 0003

0055              IF ( SUMSQ ) 75,110,75
0056        75    DX=(V*UY-U*UX)/SUMSQ
0057              X=X+DX
0058              DY=-(U*UY+V*UX)/SUMSQ
0059              Y=Y+DY
0060        78    IF ( ABS(DY)+ABS(DX)-1.0E-5 ) 100,80,80
            C
            C  STEP ITERATION COUNTER
            C
0061        80    ICT=ICT+1
0062              IF ( ICT-500 ) 60,85,85
0063        85    IF ( IFIT ) 100,90,100
0064        90    IF ( IN-5 ) 50,95,95
            C
            C  SET ERROR CODE TO 3
            C
0065        95    IER=3
0066              GO TO 20
0067       100    DO 105 L=1,NXX
0068              MT=KJ1-L+1
0069              TEMP=XCOF(MT)
0070              XCOF(MT)=COF(L)
0071       105    COF(L)=TEMP
0072              ITEMP=N
0073              N=NX
0074              NX=ITEMP
0075              IF ( IFIT ) 120,55,120
0076       110    IF ( IFIT ) 115,50,115
0077       115    X=XPR
0078              Y=YPR
0079       120    IFIT=0
0080              IF ( X ) 122,125,122
0081       122    IF ( ABS(Y)-ABS(X)*1.0E-4 ) 135,125,125
0082       125    ALPHA=X+X
0083              SUMSQ=X*X+Y*Y
0084              N=N-2
0085              GO TO 140
0086       130    X=0.
0087              NX=NX-1
0088              NXX=NXX-1
0089       135    Y=0.
0090              SUMSQ=0.
0091              ALPHA=X
0092              N=N-1
0093       140    L1=1
0094              L2=2
0095              COF(L2)=COF(L2)+ALPHA*COF(L1)
0096       145    DO 150 L=2,N
0097       150    COF(L+1)=COF(L+1)+ ALPHA*COF(L)-SUMSQ*COF(L-1)
0098       155    ROOTI(N2)=Y
0099              ROOTR(N2)=X
0100              N2=N2+1
0101              IF ( SUMSQ ) 160,165,160
0102       160    Y=-Y
```

```
FORTRAN IV G LEVEL  21                POLRT          DATE = 73097        11/11/21          PAGE 0004

 0103               SUMSQ=0.
 0104               GO TO 155
 0105          165  IF ( N ) 20,20,45
 0106               END
```

```
FORTRAN IV G LEVEL  21              POLRT           DATE = 73097        11/11/21            PAGE 0005

  *OPTIONS IN EFFECT*  ID,EBCDIC,SOURCE,NOLIST,NODECK,LOAD,NOMAP
  *OPTIONS IN EFFECT*  NAME = POLRT   , LINECNT =       56
  *STATISTICS*    SOURCE STATEMENTS =       106,PROGRAM SIZE =     2350
  *STATISTICS*  NO DIAGNOSTICS GENERATED
```

```
FORTRAN IV G LEVEL  21                  PVALUE             DATE = 73097          11/11/21                PAGE 0001

0001                SUBROUTINE PVALUE (PVAR)
              C
              C  GENERATION OF ITEM DIFFICULTY INDICES FOR EACH ITEM IN UNIVERSE
              C
0002                REAL MPCP
0003                COMMON /BLOCK1/ YBAR,YSD,MPOP,SPOP,KPOP,NDIST,INTPRT
0004                COMMON /BLOCK3/ P(150),Q(150),NSUB
0005                D1=1000.
0006                PBAR=MPCP/KPOP
0007                DO 173 I=1,KPOP
0008           173  Q(I)=0.
0009                IF ( PVAR .GT. 0. ) GO TO 66
0010                DO 65 I=1,KPOP
0011           65   P(I)=PBAR
0012                RETURN
0013           66   PSD=SQRT(PVAR)
              C
              C
0014                DO 100 IJ=1,100
0015                DO 74 I=1,KPOP
0016                CALL RANDND (R)
0017                Q(I)=R*PSD +PBAR
0018                IF ( Q(I) .LT. 0. ) Q(I)=0.
0019           74   IF ( Q(I) .GT. 1. ) Q(I)=1.
              C
              C  DETERMINE INITIAL MEAN AND VARIANCE OF GENERATED ITEM DIFFICULTY
              C  INDICES
              C
0020                SP=0.
0021                SPP=0.
0022                DO 81 I=1,KPOP
0023                PP=Q(I)
0024                SP=SP+PP
0025           81   SPP=SPP+PP*PP
0026                PVR=(SPP-SP*SP/KPCP)/KPOP
0027                CVR=SQRT(PVAR/PVR)
              C
              C  SCALE VARIANCE OF ITEM DIFFICULTY INDICES TO STANDARD
              C
0028                SP=0.
0029                SPP=0.
0030                DO 82 I=1,KPOP
0031                Q(I)=Q(I)*CVR
0032                IF ( Q(I) .GT. 1. ) Q(I)=1.
0033                PP=Q(I)
0034                SP=SP+PP
0035           82   SPP=SPP+PP*PP
0036                PVR=(SPP-SP*SP/KPCP)/KPOP
              C
              C  SCALE MEAN OF ITEM DIFFICULTY INDICES
0037                D4=SP/KPCP-PBAR
0038                SP=0.
0039                SPP=0.
0040                DO 84 I=1,KPOP
```

```
FORTRAN IV G LEVEL  21              PVALUE          DATE = 73097        11/11/21            PAGE 0002

              Q(I)=Q(I)-D4
              IF ( Q(I) .LT. 0. ) Q(I)=0.
              IF ( Q(I) .GT. 1. ) Q(I)=1.
              PP=Q(I)
              SP=SP+PP
       84     SPP=SPP+PP*PP
              PVR=(SPP-SP*SP/KPOP)/KPOP
              PBR=SP/KPOP
              D2=ABS(PVAR-PVR)
              D3=ABS(PBAR-PBR)
              I1=(D1+.0005)*1000.
              I2=(D2+.0005)*1000.
              I3=(D3+.0005)*1000.
              IF ( I2 .LE. 5 .AND. I3 .LE. 5 ) GO TO 103
              IF ( I2 .GE. I1 ) GO TO 100
              D1=D2
              DO 90 I=1,KPOP
       90     P(I)=Q(I)
      100     CONTINUE
      103     DO 104 I=1,KPOP
      104     P(I)=Q(I)
              NSTOP=KPOP-1
              DO 110 I=1,NSTOP
              JJ=I+1
              DO 110 J=JJ,KPOP
              IF ( P(I) .GE. P(J) ) GO TO 110
              TEMP=P(J)
              P(J)=P(I)
              P(I)=TEMP
      110     CONTINUE
              DO 120 I=1,KPOP
      120     Q(I)=0.
              RETURN
              END
```

```
FORTRAN IV G LEVEL  21                 PVALUE           DATE = 73097         11/11/21            PAGE 0003

  *OPTIONS IN EFFECT*  ID,EBCDIC,SOURCE,NOLIST,NODECK,LOAD,NOMAP
  *OPTIONS IN EFFECT*  NAME = PVALUE  , LINECNT =       56
  *STATISTICS*    SOURCE STATEMENTS =       74,PROGRAM SIZE =     2124
  *STATISTICS*  NO DIAGNOSTICS GENERATED
```

```
FORTRAN IV G LEVEL  21              RANDND          DATE = 73097        11/11/21              PAGE 0001

            SUBROUTINE RANDND (X)
C
C   GENERATION OF RANDOM STANDARD NORMAL DEVIATE
C
            COMMON /BLOCK5/ N,C(290)
            VTAIL=C(200)
            SGN=1.
            CALL RUNIF(U)
            IF ( U .LT. .5 ) SGN=-1.
            CALL RUNIF(U)
            IV2=1000.*U
            IV1=IV2/10
            V=100.*U-.1*FLOAT(IV2)
            IF ( U .GE. .79 ) GO TO 10
            X=(C(IV1+1)+V)*SGN
            RETURN
        10  IF ( U .GE. .94 ) GO TO 20
            X=(C(IV2-770)+V)*SGN
            RETURN
        20  IF ( U .GE. VTAIL ) GO TC 30
            J=170
        21  J=J+1
            IF ( U .GE. C(J) ) GO TO 21
            IF ( U .LT. C(J+30) ) GO TC 23
            CALL RUNIF(U)
            X=(C(J-30)+.1*U)*SGN
            RETURN
        23  CALL RUNIF(U)
            CALL RUNIF(V)
            U1=AMIN1(U,V)
            U2=AMAX1(U,V)
            IF ( U2 .GE. C(J+60) ) GO TO 25
        24  X=(C(J-30)+.1*U1)*SGN
            RETURN
        25  W=-.5*(.1*U1-.1)*(2.*C(J-30)+.1*U1+.1)
            IF ( (EXP(W)-1.)*C(J+90)-U2+U1 ) 23,23,24
        30  CALL RUNIF(U1)
            CALL RUNIF(U2)
            S=U1*U1+U2*U2
            IF ( S .GE. 1. ) GO TO 30
            T=SQRT((9.-2.*ALOG(S))/S)
            IF ( U1*T .LE. 3. ) GO TC 32
            X=U1*T*SGN
            RETURN
        32  IF ( U2*T .LT. 3. ) GO TC 30
            X=U2*T*SGN
            RETURN
            END
```

```
FORTRAN IV G LEVEL  21                RANDND           DATE = 73097         11/11/21          PAGE 0002

  *OPTIONS IN EFFECT*  ID,EBCDIC,SOURCE,NOLIST,NODECK,LOAD,NOMAP
  *OPTIONS IN EFFECT*  NAME = RANDND  , LINECNT =       56
  *STATISTICS*    SOURCE STATEMENTS =       45,PROGRAM SIZE =     1410
  *STATISTICS*  NO DIAGNOSTICS GENERATED
```

```
FORTRAN IV G LEVEL  21                 ROUND              DATE = 73097          11/11/21              PAGE 0001

 0001                  SUBROUTINE ROUND
                 C
                 C   ROUNDING SUBROUTINE FOR NORMAL AND LOGNORMAL FUNCTIONS.  AREA
                 C   BETWEEN ADJACENT INTEGER SCORES IS COMPUTED BY MEANS OF THE
                 C   TRAPEZOID FORMULA.  THE ROUNDING VALUE IS THAT CONTINUOUS TEST
                 C   SCORE SUCH THAT ONE-HALF OF THE AREA WITHIN THE SCORE INTERVAL IS
                 C   ABOVE THAT POINT.
 0002                  REAL MPOP
 0003                  COMMON /BLOCK1/ YBAR,YSD,MPOP,SPOP,KPOP,NDIST,IPRT
 0004                  COMMON /BLOCK2/ RND(150)
 0005                  DIMENSION Y(150)
                 C
 0006                  DELTA=.01
 0007                  YVAR=YSD*YSD
 0008                  VPOP=SPOP*SPOP
 0009                  PI=3.1415927
 0010                  GO TO (40,50,50), NDIST
 0011             40   CND1=1./SQRT(2.*PI*VPOP)
 0012                  CND2=2.*VPOP
 0013                  GO TO 60
 0014             50   CLN1=1./SQRT(2.*PI*YVAR)
 0015                  CLN2=2.*YVAR
 0016             60   CONTINUE
                 C
                 C
 0017                  SAREA=0.
 0018                  IF ( IPRT .EQ. 1 ) WRITE (6,7)
 0019                  DO 100 I=1,KPOP
 0020                  N=I-1
 0021                  NABSC=N
 0022                  IF ( NDIST .EQ. 3 ) NABSC=KPOP-N
                 C
                 C   TEST TO DETERMINE IF INTEGRATION IS NECESSARY
                 C
 0023                  GO TO (110,115,115), NDIST
 0024            110   Y1=CND1*EXP(-((N-MPOP)**2)/CND2)
 0025                  Y2=CND1*EXP(-((I-MPOP)**2)/CND2)
 0026                  GO TO 119
 0027            115   IF ( N .GT. 0 ) GO TO 117
 0028                  Y1=0.
 0029                  GO TO 118
 0030            117   Y1=(CLN1/N)*EXP(-((ALOG(FLOAT(N))-YBAR)**2)/CLN2)
 0031            118   Y2=(CLN1/I)*EXP(-((ALOG(FLOAT(I))-YBAR)**2)/CLN2)
 0032            119   IF ( ABS(Y1-Y2) .GT. .005 ) GO TO 120
 0033                  RND(I)=N+.5
 0034                  IF ( IPRT .EQ. 1 ) WRITE (6,5) NABSC,Y1
 0035                  SAREA=SAREA+(Y1+Y2)/2.
 0036                  GO TO 100
                 C
                 C   PERFORM INTEGRATION FROM X TO X+1
                 C
 0037            120   DO 150 J=1,101
 0038                  K=J-1
 0039                  X=N+K*DELTA
```

```
FORTRAN IV G LEVEL  21                  ROUND              DATE = 73097          11/11/21                PAGE 0002

            GO TO (130,131,131), NDIST
      130   Y(J)=CND1*EXP(-((X-MPOP)**2)/CND2)
            GO TO 150
      131   IF ( X .GT. 0. ) GO TO 135
            Y(J)=0.
            GO TC 150
      135   Y(J)=(CLN1/X)*EXP(-((ALOG(X)-YBAR)**2)/CLN2)
      150   CONTINUE
            IF (IPRT .EQ. 1) WRITE (6,5) NABSC,(Y(J),J=1,101,10)
            AREA=0.
            DC 155 J=2,100
      155   AREA=AREA+Y(J)
            AREA=DELTA*((Y(1)+Y(101))/2.+AREA)
            SAREA=SAREA+AREA
C
C  CALCULATE ROUNDING VALUE FOR SCORE INTERVAL
C
            P=C.
            DO 160 J=1,100
            K=J+1
            P=P+((Y(J)+Y(K))*DELTA/2.)/AREA
            IF ( P .LT. .5 ) GO TO 160
            RND(I)=N+J*DELTA
            GC TC 100
      160   CONTINUE
            WRITE (6,3) NABSC
            RND(I)=N+.5
      100   CCNTINUE
            IF ( IPRT .EQ. 0 ) GO TO 199
            WRITE (6,6) SAREA
            WRITE (6,1)
C
C  ARRANGE RCUNDING VALUES IN ACCORDANCE WITH NORMATIVE DISTRIBUTION
C
      199   GO TO (220,220,200), NDIST
      200   DO 210 I=1,KPOP
      210   Y(I)=KPOP-RND(I)
            DO 215 I=1,KPOP
      215   RND(I)=Y(KPOP-I+1)
      220   IF ( IPRT .EQ. 0 ) RETURN
            DO 225 I=1,KPOP
            J=I-1
      225   WRITE (6,2) J,RND(I),I
C
C
            RETURN
C
C
        1   FORMAT (///22H ROUNDING RULE   SCORE,5X,5HROUND,5X,5HSCORE//)
        2   FORMAT (17X,I5,F10.2,5X,I5)
        3   FORMAT (26H PROBLEM IN RCUND SUBR X =I5)
        5   FORMAT (1X,I3,11F8.4)
        6   FORMAT (12H APPROX AREAF15.7)
        7   FORMAT (//23H DISTRIBUTICN ORDINATES//
```

```
FORTRAN IV G LEVEL  21            ROUND          DATE = 73097        11/11/21           PAGE 0003
                    1        3X,1HN,3X,2H.0,6X,2H.1,6X,2H.2,6X,2H.3,6X,2H.4,6X,2H.5,
                    26X,2H.6,6X,2H.7,6X,2H.8,6X,2H.9,5X,3H1.0//)
 0084                END
```

```
FORTRAN IV G LEVEL  21              ROUND           DATE = 73097        11/11/21          PAGE 0004

  *OPTIONS IN EFFECT*  ID,EBCDIC,SOURCE,NOLIST,NODECK,LOAD,NOMAP
  *OPTIONS IN EFFECT*  NAME = ROUND    , LINECNT =       56
  *STATISTICS*    SOURCE STATEMENTS =        84,PROGRAM SIZE =      3616
  *STATISTICS*  NO DIAGNOSTICS GENERATED
```

```
FORTRAN IV G LEVEL  21              RUNIF          DATE = 73097        11/11/21              PAGE 0001

0001                 SUBROUTINE RUNIF(R)
               C
               C  UNIFORM RANDOM NUMBER GENERATOR (.00 - 1.00) USING POWER RESIDUE
               C  TECHNIQUE LESIGNED FOR 360 SYSTEM
               C
0002                 INTEGEF EASE
0003                 COMMON /BLOCK5/ BASE,C(290)
0004                 IY=EASE*65539
0005                 IF  ( IY) 5,6,6
0006            5    IY=IY+2147483647+1
0007            6    R=IY
0008                 R=R*.4656613E-9
0009                 BASE=IY
0010                 RETURN
0011                 END
```

```
FORTRAN IV G LEVEL  21                RUNIF           DATE = 73097         11/11/21            PAGE 0002

 *OPTIONS IN EFFECT*  ID,EBCDIC,SOURCE,NOLIST,NODECK,LOAD,NOMAP
 *OPTIONS IN EFFECT*  NAME = RUNIF    , LINECNT =       56
 *STATISTICS*    SOURCE STATEMENTS =        11,PROGRAM SIZE =      416
 *STATISTICS*  NO DIAGNOSTICS GENERATED
```

```
FORTRAN IV G LEVEL  21              SETUP             DATE = 73097        11/11/21          PAGE 0001

0001             SUBROUTINE SETUP (NI,NJ,NK,XMULT)
           C
           C  COMPUTATION OF BOUNDARIES FOR ESTIMATED CENTRAL MOMENTS:
           C  PREPARATION OF X MATRIX FOR USE BY LP SUBROUTINE
           C
           C  SUBROUTINE WRITTEN ORIGINALLY BY DR. RICHARD K. HILL,JR
           C  VIRGINIA POLYTECHNIC INSTITUTE
           C
0002             COMMON /BLOCK6/ SSUM(4),X(10,157),B(6,4),W(6)
0003             DIMENSION G(10),OBJF(4)
0004             DATA G/1H0,1H1,1H2,1H3,1H4,1H5,1H6,1H7,1H8,1H9/
0005             DATA OBJF/4HOBJE,4HCT F,4HUNCT,4HION /
0006             DATA EQU,F,BLANK/3HEQU,1HF,1H /
0007             NI1=NI+1
0008             NI2=NI+2
0009             NI3=NI+3
0010             NI4=NI+4
0011             X(1,5)=NI
0012             DO 82 J=7,NJ
0013             X(NI1,J)=F
0014             JH=(J-7)/100
0015             JT=(J-7-JH*100)/10
0016             JU=J-7-JH*100-JT*10
0017             X(NI2,J)=G(JH+1)
0018             X(NI3,J)=G(JT+1)
0019        82   X(NI4,J)=G(JU+1)
0020             DO 83 I=3,NI
0021             X(I,1)=EQU
0022             X(I,2)=BLANK
0023             JT=(I-3)/10
0024             JU=I-3-JT*10
0025             X(I,3)=G(JT+1)
0026        83   X(I,4)=G(JU+1)
0027             DO 84 I=4,NI
0028             DO 84 J=7,NJ
0029             XJ=J
0030        84   X(I,J)=(XJ-7.)**(I-3)
0031             DO 85 J=7,NJ
0032             X(3,J)=1.
0033             XJ=J
0034        85   X(1,J)=((XJ-7.)**(NK -2))*XMULT
0035             DO 86 I=4,NI
0036        86   X(I,5)=SSUM(I-3)
0037             X(3,5)=1.
0038             DO 93 J=1,4
0039        93   X(2,J)=OBJF(J)
0040             RETURN
0041             END
```

```
FORTRAN IV G LEVEL  21                    SETUP              DATE = 73097          11/11/21             PAGE 0002

  *OPTIONS IN EFFECT*  ID,EBCDIC,SOURCE,NOLIST,NODECK,LOAD,NOMAP
  *OPTIONS IN EFFECT*  NAME = SETUP    , LINECNT =       56
  *STATISTICS*    SOURCE STATEMENTS =        41,PROGRAM SIZE =      1388
  *STATISTICS*  NO DIAGNOSTICS GENERATED
```

```
FORTRAN IV G LEVEL  21            TMOMTS          DATE = 73097        11/11/21            PAGE 0001

0001              SUBROUTINE TMOMTS (TU1,TU2,TU3,TU4)
            C
            C  CALCUIATION OF 1ST MOMENT ANC 2ND THROUGH 4TH CENTRAL
            C  MOMENTS FOR TRUNCATED DISTRIEUTION.  MOMENTS ARE THOSE OBTAINED
            C  THROUGH SAMPLING 5000 VALUES FROM NORMATIVE DISTRIBUTION.
            C
0002              REAL MPCP
0003              CCMMCN /BLCCK1/ YEAR,YSD,MEOP,SPOP,KPOP,NDIST,INTPRT
0004              DIMENSION U(4)
0005              DO 100 I=1,4
0006         100  U(I)=0.
0007              DC 300 I=1,5000
0008              CALL RANDND (Z)
0009              GO TO (200,210,210), NDIST
0010         200  X=Z*SPCP+MEOP
0011              GO TC 230
0012         210  X=EXP(Z*YSD+YBAR)
0013         230  IF ( X .GT. KPOP ) X=KPOF
0014              IF ( X .LT. 0. ) X=0.
0015              DO 240 J=1,4
0016         240  U(J)=U(J)+X**J
0017         300  CONTINUE
0018              DO 350 I=1,4
0019         350  U(I)=U(I)/5000.
0020              TU1=U(1)
0021              TU2=U(2)-U(1)**2
0022              TU3=U(3)-3.*U(1)*U(2)+2.*U(1)**3
0023              TU4=U(4)-4.*U(3)*U(1)+6.*U(2)*U(1)**2-3.*U(1)**4
0024              IF ( NDIST .NE. 3 ) RETUFN
0025              TU1=KPCP-U(1)
0026              TU3=-TU3
0027              RETURN
0028              END
```

```
FORTRAN IV G LEVEL  21              TMOMTS          DATE = 73097         11/11/21           PAGE 0002

  *OPTIONS IN EFFECT*  ID,EBCDIC,SOURCE,NOLIST,NODECK,LOAD,NOMAP
  *OPTIONS IN EFFECT*  NAME = TMCMTS  , LINECNT =       56
  *STATISTICS*    SOURCE STATEMENTS =       28,PROGRAM SIZE =      1100
  *STATISTICS*  NO DIAGNOSTICS GENERATED
```

```
FORTRAN IV G LEVEL  21              TTABLE          DATE = 73097        11/11/21            PAGE 0001

0001             SUBROUTINE TTABLE (T,N,AREA)
           C
           C  CALCULATION OF STUDENT'S T CORRESPONDING TO PRESCRIBED ALPHA LEVEL
           C  (TWO-TAILED PROBABILITY)
           C
           C     ARGUMENTS
           C
           C        T = STUDENT'S T
           C        N = DEGREES OF FREEDOM ASSOCIATED WITH T
           C     AREA = RELATIVE AREA ASSOCIATED WITH CONFIDENCE INTERVAL
           C
           C  POLYNOMIAL APPROXIMATIONS USED FOR BOTH T AND Z
           C
           C  REFERENCE:  ABRAMOWITZ, M. & STEGUM, I. A. HANDBOOK OF MATHEMATICAL
           C              FUNCTIONS. NATIONAL BUREAU OF STANDARDS APPLIED
           C              MATHEMATICS SERIES NO. 55, 1964.
           C              (SEE 26.2.23 AND 26.7.5)
           C
0002             IF ( N .GT. 0 ) GO TO 100
0003             WRITE (6,1) N
0004             GO TO 150
0005         100 P=(1.-AREA)/2.
0006             IF ( P .GT. 0. .AND. P .LE. .5 ) GO TO 200
0007             WRITE (6,2) P
0008         150 T=0.
0009             RETURN
0010         200 X=SQRT(ALOG(1./(P*P)))
           C
0011             Z=X-(2.515517+.802853*X+.010328*X*X)/
               1      (1.+1.432788*X+.189269*X*X+.001308*X**3)
           C
0012             Z3=Z**3
0013             Z5=Z**5
0014             Z7=Z**7
0015             Z9=Z**9
0016             G1=(Z3+Z)/4.
0017             G2=(5.*Z5+16.*Z3+3.*Z)/96.
0018             G3=(3.*Z7+19.*Z5+17.*Z3-15.*Z)/384.
0019             G4=(79.*Z9+776.*Z7+1482.*Z5-1920.*Z3-945.*Z)/92160.
           C
0020             T=Z+G1/N+G2/N**2+G3/N**3+G4/N**4
           C
0021             RETURN
0022           1 FORMAT (//27H ERROR IN T-TABLE ARG: DF =I12//)
0023           2 FORMAT (//26H ERROR IN T-TABLE ARG: P =E15.5//)
0024             END
```

```
FORTRAN IV G LEVEL  21              TTABLE          DATE = 73097          11/11/21          PAGE 0002

  *OPTIONS IN EFFECT*  ID,EBCDIC,SOURCE,NOLIST,NODECK,LOAD,NOMAP
  *OPTIONS IN EFFECT*  NAME = TTABLE  , LINECNT =       56
  *STATISTICS*    SOURCE STATEMENTS =        24,PROGRAM SIZE =     1320
  *STATISTICS*  NO DIAGNOSTICS GENERATED
```

```
FORTRAN IV G LEVEL  21                VCOV              DATE = 73097          11/11/21              PAGE 0001

 0001                  SUBROUTINE VCOV (XX,NSPT,IPT,VCOVIT)
                 C
                 C  COMPUTATION OF VARIANCE OF ITEM-TOTAL COVARIANCES FOR ONE SUBTEST
                 C
 0002                  DIMENSION XX(100,150),T(150)
 0003                  SUMT=0.
 0004                  DO 100 I=1,NSPT
 0005                  T(I)=0.
 0006                  DO 150 J=1,IPT
 0007             150  T(I)=T(I)+XX(I,J)
 0008                  T(I)=T(I)/IPT
 0009             100  SUMT=SUMT+T(I)
 0010                  SUMC=0.
 0011                  SUMC2=0.
 0012                  DO 200 I=1,IPT
 0013                  COV=0.
 0014                  TEMP=0.
 0015                  DO 250 J=1,NSPT
 0016                  TEMP=TEMP+XX(J,I)
 0017             250  COV=COV+XX(J,I)*T(J)
 0018                  COV=COV/NSPT-TEMP*SUMT/NSPT**2
 0019                  SUMC=SUMC+COV
 0020             200  SUMC2=SUMC2+COV*COV
 0021                  VCOVIT=SUMC2/IPT-SUMC*SUMC/IPT**2
 0022                  RETURN
 0023                  END
```

```
FORTRAN IV G LEVEL  21                VCOV              DATE = 73097          11/11/21             PAGE 0002

  *OPTIONS IN EFFECT*  ID,EBCDIC,SOURCE,NOLIST,NODECK,LOAD,NOMAP
  *OPTIONS IN EFFECT*  NAME = VCOV    , LINECNT =       56
  *STATISTICS*    SOURCE STATEMENTS =       23,PROGRAM SIZE =      1502
  *STATISTICS*  NO DIAGNOSTICS GENERATED

  *STATISTICS*  NO DIAGNOSTICS THIS STEP 2
```

```
IEF285I   SYS1.FORTRAN                                      PASSED
IEF285I   VOL SER NOS= SYSTM3.
IEF285I   SYS73097.T111053.RV000.AGB005Z2.LOADSET           PASSED
IEF285I   VOL SER NOS= SYSDA1.
IEF285I   SYS73097.T111053.SV000.AGB005Z2.R0000001          SYSOUT
IEF285I   VOL SER NOS= SYSDA4.
IEF285I   SYS73097.T111053.RV000.AGB005Z2.S0000002          SYSIN
IEF285I   VOL SER NOS= SYSDA3.
IEF285I   SYS73097.T111053.RV000.AGB005Z2.S0000002          DELETED
IEF285I   VOL SER NOS= SYSDA3.
CCN011I STEP RETURN CODE =    0

CCN301I   SYSPRINT PRINTED    88 TOTAL PAGES, REQUIRED    50 TRACKS.

  REGION ALLOCATION       CORE USED          MAIN STORAGE       EXTENDED STORAGE
        100K                 98K                100K                   0K

   STEP CPU TIME          JOB CPU TIME       STEP I/O COUNT       JOB I/O COUNT
       19.66S                19.66S              622                  622

XXLKED   EXEC  PGM=IEWL,REGION=&RL,TIME=&TL,COND=(5,LT,FORT),PARM=MAP   00000090
IEF653I SUBSTITUTION JCL - PGM=IEWL,REGION=154K,TIME=1439,COND=(5,LT,FORT),PARM=MAP
XXSYSLIB DD DISP=(SHR,PASS),DSN=$YS&LEVEL..FORTLIB                      00000100
IEF653I SUBSTITUTION JCL - DISP=(SHR,PASS),DSN=$YS1.FORTLIB
XXSYSLIN DD DSN=&&LOADSET,DISP=(OLD,DELETE)                             00000110
XX       DD    DDNAME=SYSIN                                             00000120
XXSYSLMOD DD DISP=(MOD,PASS),DSN=&&GOSET(GOFORTG),UNIT=SYSDA,           00000130
XX SPACE=(3072,(20,20,1),RLSE)                                          00000140
XXSYSUT1 DD UNIT=SYSDA,SPACE=(TRK,(20,20))                              00000150
XXSYSPRINT DD SYSOUT=A,SPACE=(TRK,&PL,RLSE),DCB=BLKSIZE=2178            00000160
IEF653I SUBSTITUTION JCL - SYSOUT=A,SPACE=(TRK,10,RLSE),DCB=BLKSIZE=2178
IEF236I ALLOC. FOR AGB005Z2 LKED
IEF237I 1C0   ALLOCATED TO SYSLIB
IEF237I 230   ALLOCATED TO SYSLIN
IEF237I 230   ALLOCATED TO SYSLMOD
IEF237I 230   ALLOCATED TO SYSUT1
IEF237I 630   ALLOCATED TO SYSPRINT
```

F128-LEVEL LINKAGE EDITOR OPTIONS SPECIFIED MAP
DEFAULT OPTION(S) USED - SIZE=(151552,40960)

MODULE MAP

CONTROL SECTION			ENTRY							
NAME	ORIGIN	LENGTH	NAME	LOCATION	NAME	LOCATION	NAME	LOCATION	NAME	LOCATION
MAIN	00	2030								
ALLOC	2030	A62								
BLOCK5	2A98	48C								
BLOCK8	2F28	1340								
BMOMTS	4268	BC0								
DATA	4D68	54A								
ESTORS	52B8	33A8								
FREAKS	8660	338								
FRKOUT	8998	588								
FRV	8F20	2E0								
GAMMA	9200	40C								
LP	9610	AF0								
MOMVAL	A100	470								
NEGHGR	A570	738								
PEARSN	ACA8	2900								
POOL	D5A8	FB78								
POIRT	1D120	92E								
PVALUE	1DA50	84C								
RANDND	1E2A0	582								
ROUND	1E828	E20								
RUNIF	1F648	1A0								
SETUP	1F7E8	56C								
TMOMTS	1FD58	44C								
TTABLE	201A8	528								
VCOV	206D0	5DE								
IHCSLOG *	20CB0	1B6	ALOG10	20CB0	ALOG	20CC8				
IHCSATN2*	20E68	1CB	ATAN2	20E68	ATAN	20E7C				
IHCSSCN *	21038	1D9	COS	21038	SIN	21050				
IHCFEXIT*	21218	1C	EXIT	21218						
IHCSEXP *	21238	192	EXP	21238						
IHCFRXPI*	213D0	141	FRXPI#	213D0						
IHCFRXPR*	21518	183	FRXPR#	21518						
IHCECOMH*	216A0	F61	IBCOM#	216A0	FDIOCS#	2175C	INTSWTCH	225E6		
IHCCOMH2*	22608	65D	SEQDASD	22980						
IHCSSQRT*	22C68	145								

```
 NAME       ORIGIN  LENGTH        NAME      LOCATION    NAME      LOCATION    NAME      LOCATION    NAME      LOCATION
                                  SQRT      22C68
 IHCFCVTH*  22DB0    119D
                                  ADCON#    22DB0       FCVAOUTP  22E5A       FCVLOUTP  22EEA       FCVZOUTP  2303A
                                  FCVIOUTP  233E8       FCVEOUTP  238EA       FCVCOUTP  23B04       INT6SWCH  23DEB
 IHCFMAXR*  23F50      C9
                                  MAX1      23F50       MIN1      23F66       AMAX1     23F7C       AMIN1     23F92
 IHCEFNTH*  24020     542
                                  ARITH#    24020       ADJSWTCH  243BC
 IHCEFIOS*  24568     EBC
                                  FIOCS#    24568       FIOCSBEP  2456E
 IHCFIOS2*  25428     52E
 IHCEFRM *  25958     5D4
                                  ERRMCN    25958       IHCERRE   25970
 IHCUOPT *  25F30     340
 IHCETRCH*  26270     28E
                                  IHCTRCH   26270       ERRTRA    26278
 IHCUATBL*  26500     638
$BLANKCCM   26B38    1C20
 BLCCK1     28758      1C
 BLOCK2     28778     258
 BLCCK3     289D0     4B4
 BLCCK7     28E88     B98
 BLOCK6     29A20    1910

 ENTRY ADDRESS         0C
 TOTAL LENGTH       2B330

****GOFORTG   DOES NOT EXIST BUT HAS BEEN ADDEL TO DATA SET
```

```
IEF285I   SYS1.FORTLIB                                         PASSED
IEF285I   VOL SER NOS= DRUM01.
IEF285I   SYS73097.T111053.RV000.AGB005Z2.LOADSET              DELETED
IEF285I   VOL SER NOS= SYSDA1.
IEF285I   SYS73097.T111053.RV000.AGB005Z2.GOSET                PASSED
IEF285I   VOL SER NOS= SYSDA1.
IEF285I   SYS73097.T111053.RV000.AGB005Z2.R0000003             DELETED
IEF285I   VOL SER NOS= SYSDA1.
IEF285I   SYS73097.T111053.SV000.AGB005Z2.R0000004             SYSOUT
IEF285I   VOL SER NOS= SYSDA4.
CCN011I STEP RETURN CODE =    0

CCN301I   SYSPRINT PRINTED     2 TOTAL PAGES, REQUIRED     2 TRACKS.

  REGION ALLOCATION        CORE USED         MAIN STORAGE      EXTENDED STORAGE
        154K                 150K               154K                 0K

   STEP CPU TIME          JOB CPU TIME      STEP I/O COUNT     JOB I/O COUNT
       0.95S                 20.61S             458                1080

XXGO      EXEC  PGM=*.LKED.SYSLMOD,COND=((5,LT,FORT),(5,LT,LKED)),        00000170
XX   REGION=&RG,TIME=&TG                                                   00000180
IEF653I SUBSTITUTION JCL - REGION=200K,TIME=1439
XXFT05F001 DD DDNAME=SYSIN                                                 00000190
XXSTEPLIB  DD DISP=(SHR,PASS),DSN=$YS&LEVEL..FORTRAN                       00000200
IEF653I SUBSTITUTION JCL - DISP=(SHR,PASS),DSN=$YS1.FORTRAN
XXFT06F001 DD SYSOUT=A,DCB=(BLKSIZE=3458,RECFM=FBA,LRECL=133),             00000210
XX   SPACE=(TRK,&PG,RLSE)                                                  00000220
IEF653I SUBSTITUTION JCL - SPACE=(TRK,50,RLSE)
//GO.SYSIN DD *
//
IEF236I ALLOC. FOR AGB005Z2 GO
IEF237I 230    ALLOCATED TO PGM=*.DD
IEF237I 230    ALLOCATED TO FT05F001
IEF237I 342    ALLOCATED TO STEPLIB
IEF237I 630    ALLOCATED TO FT06F001


PROBLEM NO.   5247

PARAMETERS INPUTTED     ITEM-EXAMINEE SAMPLING PLAN     SWITCHES

 MEAN         20.000     NO. OF SUBTESTS           4     ITEM SAMPLING PLAN       2
 VAR          36.364     ITEMS PER SUBTEST        10     RELATIVE FREQ DIST       0
 K            40         EXAMINEES PER SUBTEST    50     NORMATIVE DIST           1
 KR-20         0.800     NO. OF REPLICATIONS       5     INTERMEDIATE PRINTOUT    1
 VAR(I)        0.050
 THETA         0.100
```

ITEM DIFFICULTY INDICES

0.983	0.885	0.746	0.743	0.719	0.710	0.699	0.698	0.688	0.679
0.671	0.650	0.640	0.621	0.610	0.604	0.597	0.597	0.586	0.575
0.554	0.523	0.512	0.485	0.457	0.424	0.398	0.362	0.326	0.321
0.308	0.306	0.301	0.217	0.197	0.172	0.161	0.142	0.106	0.025

DISTRIBUTION ORDINATES

N	.0	.1	.2	.3	.4	.5	.6	.7	.8	.9	1.0
0	0.0003										
1	0.0005										
2	0.0008										
3	0.0012										
4	0.0020										
5	0.0030										
6	0.0045										
7	0.0065										
8	0.0091										
9	0.0125										
10	0.0167										
11	0.0217	0.0223	0.0228	0.0234	0.0239	0.0245	0.0251	0.0257	0.0262	0.0268	0.0274
12	0.0274	0.0280	0.0287	0.0293	0.0299	0.0305	0.0312	0.0318	0.0324	0.0331	0.0337
13	0.0337	0.0344	0.0350	0.0357	0.0363	0.0370	0.0377	0.0383	0.0390	0.0397	0.0403
14	0.0403	0.0410	0.0417	0.0423	0.0430	0.0436	0.0443	0.0450	0.0456	0.0463	0.0469
15	0.0469	0.0476	0.0482	0.0488	0.0495	0.0501	0.0507	0.0513	0.0519	0.0525	0.0531
16	0.0531	0.0537	0.0542	0.0548	0.0554	0.0559	0.0564	0.0570	0.0575	0.0580	0.0585
17	0.0585										
18	0.0626										
19	0.0653										
20	0.0662										
21	0.0653										
22	0.0626										
23	0.0585	0.0580	0.0575	0.0570	0.0564	0.0559	0.0554	0.0548	0.0542	0.0537	0.0531
24	0.0531	0.0525	0.0519	0.0513	0.0507	0.0501	0.0495	0.0488	0.0482	0.0476	0.0469
25	0.0469	0.0463	0.0456	0.0450	0.0443	0.0436	0.0430	0.0423	0.0417	0.0410	0.0403
26	0.0403	0.0397	0.0390	0.0383	0.0377	0.0370	0.0363	0.0357	0.0350	0.0344	0.0337
27	0.0337	0.0331	0.0324	0.0318	0.0312	0.0305	0.0299	0.0293	0.0287	0.0280	0.0274
28	0.0274	0.0268	0.0262	0.0257	0.0251	0.0245	0.0239	0.0234	0.0228	0.0223	0.0217
29	0.0217										
30	0.0167										
31	0.0125										
32	0.0091										
33	0.0065										
34	0.0045										
35	0.0030										
36	0.0020										
37	0.0012										
38	0.0008										
39	0.0005										

APPROX AREA 0.9991509

ROUNDING RULE

SCORE	ROUND	SCORE
0	0.50	1
1	1.50	2
2	2.50	3
3	3.50	4
4	4.50	5
5	5.50	6
6	6.50	7
7	7.50	8
8	8.50	9
9	9.50	10
10	10.50	11
11	11.53	12
12	12.53	13
13	13.53	14
14	14.52	15
15	15.52	16
16	16.52	17
17	17.50	18
18	18.50	19
19	19.50	20
20	20.50	21
21	21.50	22
22	22.50	23
23	23.49	24
24	24.49	25
25	25.49	26
26	26.48	27
27	27.48	28
28	28.48	29
29	29.50	30
30	30.50	31
31	31.50	32
32	32.50	33
33	33.50	34
34	34.50	35
35	35.50	36
36	36.50	37
37	37.50	38
38	38.50	39
39	39.50	40

MOMENT	THEORETICAL DIST	TRUNCATED DIST
U1'	0.2000000E 02	0.1990305E 02
U2	0.3636362E 02	0.3607422E 02
U3	0.0	-0.7273438E 01
U4	0.3966938E 04	0.3946250E 04

SUBTEST	ITEMS									
1	23	4	12	33	5	17	3	2	27	31
2	22	18	35	6	10	28	16	40	20	15
3	9	8	26	34	39	7	11	1	32	21
4	13	14	19	24	25	29	30	36	37	38

SUBTEST	MEAN	VARIANCE	U3	U4	VAR(I)	VAR(E)	VAR(IE)	THETA'
	ESTIMATES OF MOMENTS				COMPONENTS OF VARIANCE			
1	0.21840E 02	0.23850E 02	0.95852E 02	0.17583E 04	0.60998E-01	0.76191E-02	0.18589E 00	0.41629E-01
2	0.19440E 02	0.23792E 02	-0.12930E 03	0.20343E 04	0.47138E-01	0.90975E-02	0.19886E 00	0.46412E-01
3	0.21120E 02	0.26655E 02	-0.14828E 03	0.10776E 03	0.76961E-01	0.98957E-02	0.17059E 00	0.58727E-01
4	0.14880E 02	0.65272E 02	0.49785E 02	0.96888E 04	0.31846E-01	0.37442E-01	0.16860E 00	0.22355E 00
MEAN	0.19320E 02	0.34892E 02	-0.32986E 02	0.33973E 04	0.54236E-01	0.16014E-01	0.18099E 00	0.92579E-01
SE	0.15631E 01	0.10149E 02	0.61927E 02	0.21399E 04	0.96344E-02	0.71584E-02	0.71004E-02	0.43804E-01

VAR(U1)-HLN 0.23118
VAR(U2)-HLN 28.49794

	MEAN	VARIANCE	U3	U4	VAR(I)	VAR(E)	VAR(IE)	THETA'
90.0 PERCENT CONFIDENCE INTERVAL								
FROM	0.15642E 02	0.11011E 02	-0.17871E 03	-0.16382E 04	0.31565E-01	-0.83117E-03	0.16428E 00	-0.10499E-01
TO	0.22998E 02	0.58773E 02	0.11274E 03	0.84327E 04	0.76907E-01	0.32858E-01	0.19769E 00	0.19566E 00
95.0 PERCENT CONFIDENCE INTERVAL								
FROM	0.14350E 02	0.26207E 01	-0.22991E 03	-0.34074E 04	0.23599E-01	-0.67495E-02	0.15841E 00	-0.46715E-01
TO	0.24290E 02	0.67164E 02	0.16394E 03	0.10202E 05	0.84872E-01	0.38777E-01	0.20356E 00	0.23187E 00
97.5 PERCENT CONFIDENCE INTERVAL								
FROM	0.12808E 02	-0.73854E 01	-0.29097E 03	-0.55172E 04	0.14100E-01	-0.13807E-01	0.15141E 00	-0.89904E-01
TO	0.25832E 02	0.77170E 02	0.22500E 03	0.12312E 05	0.94372E-01	0.45835E-01	0.21057E 00	0.27506E 00
99.0 PERCENT CONFIDENCE INTERVAL								
FROM	0.10258E 02	-0.23942E 02	-0.39200E 03	-0.90082E 04	-0.16179E-02	-0.25486E-01	0.13982E 00	-0.16137E 00
TO	0.28382E 02	0.93727E 02	0.32603E 03	0.15803E 05	0.11009E 00	0.57513E-01	0.22215E 00	0.34652E 00

REPLICATION NO. 1

PARAMETER	POOLED ESTIMATE	ESTIMATED SE OF ESTIMATE
MEAN	19.31999	1.56308
VAR	34.89232	10.14859
U3	-32.98643	61.92746
U4	3397.27930	2139.87842
VAR(I)	0.05424	0.00963
THETA	0.09258	0.04380
RELIABILITY OF TOTAL	0.78738	

SUBTEST	ITEMS									
1	35	40	8	10	29	11	3	31	5	32
2	33	26	4	17	1	18	34	38	19	14
3	30	39	9	28	7	2	12	13	22	36
4	6	15	16	20	21	23	24	25	27	37

SUBTEST	ESTIMATES OF MOMENTS: MEAN	VARIANCE	U3	U4	COMPONENTS OF VARIANCE: VAR(I)	VAR(E)	VAR(IE)	THETA'
1	0.18720E 02	0.25387E 02	0.13855E 03	0.36176E 04	0.68109E-01	0.10576E-01	0.17767E 00	0.60253E-01
2	0.21680E 02	0.37718E 02	-0.28021E 03	0.37062E 04	0.49887E-01	0.16372E-01	0.18767E 00	0.88091E-01
3	0.20960E 02	0.34537E 02	0.80622E 02	0.56046E 04	0.47510E-01	0.14857E-01	0.19249E 00	0.77992E-01
4	0.19760E 02	0.69068E 02	-0.97264E 02	0.10705E 05	0.22803E-01	0.37152E-01	0.19342E 00	0.19341E 00
MEAN	0.20280E 02	0.41678E 02	-0.39573E 02	0.59084E 04	0.47077E-01	0.19739E-01	0.18781E 00	0.10494E 00
SE	0.65362E 00	0.94968E 01	0.94608E 02	0.16634E 04	0.93080E-02	0.59326E-02	0.36074E-02	0.30047E-01

VAR(U1)-HLN 0.27195
VAR(U2)-HLN 121.18713

	MEAN	VARIANCE	U3	U4	VAR(I)	VAR(E)	VAR(IE)	THETA'
90.0 PERCENT CONFIDENCE INTERVAL								
FROM	0.18742E 02	0.19330E 02	-0.26220E 03	0.19943E 04	0.25174E-01	0.57791E-02	0.17932E 00	0.34231E-01
TO	0.21818E 02	0.64025E 02	0.18305E 03	0.98226E 04	0.68980E-01	0.33699E-01	0.19630E 00	0.17564E 00
95.0 PERCENT CONFIDENCE INTERVAL								
FROM	0.18202E 02	0.11479E 02	-0.34042E 03	0.61911E 03	0.17479E-01	0.87420E-03	0.17634E 00	0.93889E-02
TO	0.22358E 02	0.71877E 02	0.26127E 03	0.11198E 05	0.76676E-01	0.38604E-01	0.19928E 00	0.20048E 00
97.5 PERCENT CONFIDENCE INTERVAL								
FROM	0.17557E 02	0.21151E 01	-0.43370E 03	-0.10209E 04	0.83013E-02	-0.49750E-02	0.17278E 00	-0.20236E-01
TO	0.23003E 02	0.81240E 02	0.35455E 03	0.12838E 05	0.85853E-01	0.44454E-01	0.20284E 00	0.23011E 00
99.0 PERCENT CONFIDENCE INTERVAL								
FROM	0.16491E 02	-0.13378E 02	-0.58804E 03	-0.37345E 04	-0.68840E-02	-0.14654E-01	0.16690E 00	-0.69256E-01
TO	0.24069E 02	0.96734E 02	0.50890E 03	0.15551E 05	0.10104E 00	0.54132E-01	0.20873E 00	0.27913E 00

REPLICATION NO. 2

PARAMETER	POOLED ESTIMATE	ESTIMATED SE OF ESTIMATE
MEAN	20.27998	0.65362
VAR	41.67764	9.49681
U3	-39.57304	94.60808
U4	5908.44141	1663.35498
VAR(I)	0.04708	0.00931
THETA	0.10494	0.03005
RELIABILITY OF TOTAL	0.80760	

SUBTEST	ITEMS									
1	31	27	3	18	2	8	34	9	36	12
2	33	16	35	4	6	1	10	19	26	40
3	5	11	28	38	17	14	22	24	21	30
4	7	13	15	20	23	25	29	32	37	39

SUBTEST	ESTIMATES OF MOMENTS: MEAN	VARIANCE	U3	U4	COMPONENTS OF VARIANCE: VAR(I)	VAR(E)	VAR(IE)	THETA'
1	0.19760E 02	0.34721E 02	0.27505E 03	0.78352E 04	0.81034E-01	0.15819E-01	0.16185E 00	0.98634E-01
2	0.20480E 02	0.10257E 02	-0.44458E 02	-0.47858E 03	0.90059E-01	0.90706E-04	0.16905E 00	0.99460E-03
3	0.20400E 02	0.43192E 02	-0.24132E 03	0.68396E 04	0.34440E-01	0.20599E-01	0.19912E 00	0.10438E 00
4	0.16560E 02	0.49874E 02	-0.21782E 03	0.27035E 04	0.55701E-01	0.27048E-01	0.16630E 00	0.16384E 00
MEAN	0.19300E 02	0.34511E 02	-0.57136E 02	0.42249E 04	0.65308E-01	0.15889E-01	0.17408E 00	0.91962E-01
SE	0.92745E 00	0.86587E 01	0.11911E 03	0.19216E 04	0.12600E-01	0.57467E-02	0.84758E-02	0.33715E-01

VAR(U1)-HLN 0.21614
VAR(U2)-HLN 53.68489

	MEAN	VARIANCE	U3	U4	VAR(I)	VAR(E)	VAR(IE)	THETA'
90.0 PERCENT CONFIDENCE INTERVAL								
FROM	0.17118E 02	0.14136E 02	-0.33743E 03	-0.29677E 03	0.35659E-01	0.23662E-02	0.15414E 00	0.12626E-01
TO	0.21482E 02	0.54886E 02	0.22315E 03	0.87467E 04	0.94957E-01	0.29412E-01	0.19403E 00	0.17130E 00
95.0 PERCENT CONFIDENCE INTERVAL								
FROM	0.16351E 02	0.69771E 01	-0.43590E 03	-0.18855E 04	0.25242E-01	-0.23850E-02	0.14713E 00	-0.15249E-01
TO	0.22249E 02	0.62045E 02	0.32163E 03	0.10335E 05	0.10537E 00	0.34163E-01	0.20103E 00	0.19917E 00
97.5 PERCENT CONFIDENCE INTERVAL								
FROM	0.15436E 02	-0.15601E 01	-0.55334E 03	-0.37800E 04	0.12819E-01	-0.80510E-02	0.13877E 00	-0.48490E-01
TO	0.23164E 02	0.70582E 02	0.43907E 03	0.12230E 05	0.11780E 00	0.39829E-01	0.20939E 00	0.23241E 00
99.0 PERCENT CONFIDENCE INTERVAL								
FROM	0.13923E 02	-0.15686E 02	-0.74767E 03	-0.69149E 04	-0.77363E-02	-0.17426E-01	0.12494E 00	-0.10349E 00
TO	0.24677E 02	0.84708E 02	0.63340E 03	0.15365E 05	0.13835E 00	0.49204E-01	0.22322E 00	0.28742E 00

REPLICATION NO. 3

PARAMETER	POOLED ESTIMATE	ESTIMATED SE OF ESTIMATE
MEAN	19.29999	0.92745
VAR	34.51109	8.65873
U3	-57.13570	119.11263
U4	4224.94531	1921.56030
VAR(I)	0.06531	0.01260
THETA	0.09196	0.03371
RELIABILITY OF TOTAL	0.78625	

SUBTEST	ITEMS									
1	36	11	23	35	9	18	32	26	33	4
2	7	31	24	30	8	20	38	16	2	25
3	17	14	13	37	1	22	12	27	15	3
4	5	6	10	19	21	28	29	34	39	40

SUBTEST	ESTIMATES OF MOMENTS: MEAN	VARIANCE	U3	U4	COMPONENTS OF VARIANCE: VAR(I)	VAR(E)	VAR(IE)	THETA'
1	0.19520E 02	0.44854E 02	0.42060E 03	0.10575E 05	0.30304E-01	0.22177E-01	0.20125E 00	0.11115E 00
2	0.19920E 02	0.30753E 02	-0.22170E 03	0.30414E 04	0.49605E-01	0.13161E-01	0.19284E 00	0.69015E-01
3	0.23840E 02	0.33135E 02	0.56420E 02	0.18989E 04	0.47900E-01	0.11964E-01	0.18632E 00	0.64958E-01
4	0.17120E 02	0.31108E 02	-0.16549E 03	0.13492E 04	0.60045E-01	0.15030E-01	0.17640E 00	0.86045E-01
MEAN	0.20100E 02	0.34962E 02	0.22460E 02	0.42161E 04	0.46964E-01	0.15583E-01	0.18920E 00	0.82793E-01
SE	0.13915E 01	0.33385E 01	0.14566E 03	0.21487E 04	0.61681E-02	0.22867E-02	0.52486E-02	0.10499E-01

VAR(U1)-HLN 0.23990
VAR(U2)-HLN 62.88326

	MEAN	VARIANCE	U3	U4	VAR(I)	VAR(E)	VAR(IE)	THETA'
90.0 PERCENT CONFIDENCE INTERVAL								
FROM	0.16825E 02	0.27106E 02	-0.32030E 03	-0.84018E 03	0.32449E-01	0.10202E-01	0.17685E 00	0.58087E-01
TO	0.23374E 02	0.42818E 02	0.36522E 03	0.92725E 04	0.61478E-01	0.20964E-01	0.20155E 00	0.10750E 00
95.0 PERCENT CONFIDENCE INTERVAL								
FROM	0.15675E 02	0.24346E 02	-0.44073E 03	-0.26167E 04	0.27350E-01	0.83111E-02	0.17251E 00	0.49407E-01
TO	0.24525E 02	0.45579E 02	0.48565E 03	0.11049E 05	0.66578E-01	0.22854E-01	0.20589E 00	0.11618E 00
97.5 PERCENT CONFIDENCE INTERVAL								
FROM	0.14303E 02	0.21055E 02	-0.58435E 03	-0.47353E 04	0.21268E-01	0.60565E-02	0.16734E 00	0.39056E-01
TO	0.25897E 02	0.48870E 02	0.62927E 03	0.13168E 05	0.72659E-01	0.25109E-01	0.21107E 00	0.12653E 00
99.0 PERCENT CONFIDENCE INTERVAL								
FROM	0.12033E 02	0.15608E 02	-0.82199E 03	-0.82408E 04	0.11206E-01	0.23259E-02	0.15878E 00	0.21927E-01
TO	0.28167E 02	0.54317E 02	0.86691E 03	0.16673E 05	0.82722E-01	0.28840E-01	0.21963E 00	0.14366E 00

REPLICATION NO. 4

PARAMETER	POOLED ESTIMATE	ESTIMATED SE OF ESTIMATE
MEAN	20.09999	1.39154
VAR	34.96243	3.33849
U3	22.45996	145.66196
U4	4216.14063	2148.74561
VAR(I)	0.04696	0.00617
THETA	0.08279	0.01050
RELIABILITY OF TOTAL	0.76807	

SUBTEST	ITEMS									
1	23	31	16	26	8	17	34	32	5	24
2	2	37	20	27	11	15	21	4	25	7
3	3	14	13	19	12	29	33	18	30	22
4	1	6	9	10	28	35	36	38	39	40

SUBTEST	MEAN	VARIANCE	U3	U4	VAR(I)	VAR(E)	VAR(IE)	THETA'
	ESTIMATES OF MOMENTS				COMPONENTS OF VARIANCE			
1	0.19520E 02	0.44854E 02	0.42060E 03	0.10575E 05	0.30304E-01	0.22177E-01	0.20125E 00	0.11115E 00
2	0.19920E 02	0.30753E 02	-0.22170E 03	0.30414E 04	0.49605E-01	0.13161E-01	0.19284E 00	0.69015E-01

SUBTEST	MEAN	VARIANCE	U3	U4	VAR(I)	VAR(E)	VAR(IE)	THETA'
	ESTIMATES OF MOMENTS				COMPONENTS OF VARIANCE			
1	0.18560E 02	0.51762E 02	0.25921E 02	0.92433E 04	0.36816E-01	0.27084E-01	0.18941E 00	0.14410E 00
2	0.22720E 02	0.25044E 02	-0.86973E 02	-0.31800E 03	0.39882E-01	0.77007E-02	0.20234E 00	0.38687E-01
3	0.21120E 02	0.41059E 02	0.22391E 02	0.82256E 04	0.25451E-01	0.18776E-01	0.20833E 00	0.90992E-01
4	0.16560E 02	0.35418E 02	0.25772E 03	0.32620E 04	0.10223E 00	0.18122E-01	0.13310E 00	0.13723E 00
MEAN	0.19740E 02	0.38321E 02	0.54766E 02	0.51032E 04	0.51095E-01	0.17921E-01	0.18329E 00	0.10275E 00
SE	0.13629E 01	0.55742E 01	0.72550E 02	0.22299E 04	0.17326E-01	0.39706E-02	0.17190E-01	0.24395E-01

VAR(U1)-HLN 0.25069
VAR(U2)-HLN 90.38235

	MEAN	VARIANCE	U3	U4	VAR(I)	VAR(E)	VAR(IE)	THETA'
90.0 PERCENT CONFIDENCE INTERVAL								
FROM	0.16533E 02	0.25204E 02	-0.11596E 03	-0.14406E 03	0.10325E-01	0.85773E-02	0.14284E 00	0.45349E-01
TO	0.22947E 02	0.51437E 02	0.22549E 03	0.10351E 05	0.91865E-01	0.27264E-01	0.22374E 00	0.16016E 00
95.0 PERCENT CONFIDENCE INTERVAL								
FROM	0.15406E 02	0.20595E 02	-0.17594E 03	-0.19877E 04	-0.39989E-02	0.52946E-02	0.12863E 00	0.25180E-01
TO	0.24074E 02	0.56046E 02	0.28547E 03	0.12194E 05	0.10619E 00	0.30547E-01	0.23796E 00	0.18033E 00
97.5 PERCENT CONFIDENCE INTERVAL								
FROM	0.14062E 02	0.15099E 02	-0.24747E 03	-0.41863E 04	-0.21081E-01	0.13798E-02	0.11168E 00	0.11281E-02
TO	0.25418E 02	0.61542E 02	0.35700E 03	0.14393E 05	0.12327E 00	0.34462E-01	0.25490E 00	0.20438E 00
99.0 PERCENT CONFIDENCE INTERVAL								
FROM	0.11839E 02	0.60055E 01	-0.36583E 03	-0.78242E 04	-0.49347E-01	-0.50979E-02	0.63639E-01	-0.38670E-01
TO	0.27641E 02	0.70636E 02	0.47536E 03	0.18031E 05	0.15154E 00	0.40939E-01	0.28295E 00	0.24418E 00

REPLICATION NO. 5

PARAMETER	POOLED ESTIMATE	ESTIMATED SE OF ESTIMATE
MEAN	19.73997	1.36289
VAR	38.32063	5.57416
U3	54.76582	72.55000
U4	5103.23438	2229.90308
VAR(I)	0.05110	0.01733
THETA	0.10275	0.02439
RELIABILITY OF TOTAL	0.80431	

```
PROBLEM NO.   5247

PARAMETERS INPUTTED      ITEM-EXAMINEE SAMPLING PLAN     SWITCHES

 MEAN          20.000     NO. OF SUBTESTS           4     ITEM SAMPLING PLAN       2
 VAR           36.364     ITEMS PER SUBTEST        10     RELATIVE FREQ DIST       0
 K             40         EXAMINEES PER SUBTEST    50     NORMATIVE DIST           1
 KR-20          0.800     NO. OF REPLICATIONS       5     INTERMEDIATE PRINTOUT    1
 VAR(I)         0.050
 THETA          0.103


MOMENT           THEORETICAL DIST          TRUNCATED DIST

U1'                0.2000000E 02            0.1990305E 02
U2                 0.3636362E 02            0.3607422E 02
U3                 0.0                     -0.7273438E 01
U4                 0.3966938E 04            0.3946250E 04


                               SUMMARY (RESULTS POOLED OVER REPLICATIONS)

     PARAMETER            MEAN ESTIMATE          SE(REPLICATIONS)      MEAN SE(JACKKNIFE)        SD(SE(JACKKNIFE))

     MEAN                    19.7479858             0.3977476             1.1797161                0.3363922
     VAR                     36.8728180             2.7686863             7.4433498                2.5828648
     U3                     -10.4938774            42.1230316            98.7719116               30.5776215
     U4                    4570.0078125           859.6928711          2020.6875000              205.8659668
     VAR(I)                   0.0529360             0.0067550             0.0110072                0.0037589
     THETA                    0.0950044             0.0080368             0.0284919                0.0109973
     RELIABILITY OF TOTAL     0.7907223             0.0142391
     SE(U1')-HLN              0.4915359             0.0190267
     SE(U2)-HLN               8.2221375             1.9296589

ITEM DIFFICULTY INDICES FOR ALL DATA GENERATED


0.983  0.885  0.746  0.743  0.719  0.710  0.699  0.698  0.688  0.679
0.671  0.650  0.640  0.621  0.610  0.604  0.597  0.597  0.586  0.575
0.554  0.523  0.512  0.485  0.458  0.425  0.398  0.363  0.325  0.321
0.308  0.298  0.248  0.199  0.168  0.145  0.120  0.093  0.076  0.025
ALL INPUT PROCESSED
```

References

REFERENCES FOR CHAPTER 1

Cook, D.L. and Stufflebeam, D.L. "Estimating Test Norms from Variable Size Item and Examinee Samples." *Educational and Psychological Measurement* 27 (1967): 601-10.

Das, R.S. "Some Models for Assessment of Intelligence and Scholastic Attainment." *Journal of Psychological Research* 11 (1967): 77-90.

Hooke, R. "Symmetric Functions of a Two-way Array." *Annals of Mathematical Statistics* 27 (1956): 55-79.

Huck, S.W. and Bowers, N.D. "Item Difficulty Level and Sequence Effects in Multiple-Choice Achievement Tests." *Journal of Educational Measurement* 9 (1972): 105-11.

Johnson, M.C. and Lord, F.M. "An Empirical Study of the Stability of a Group Mean in Relation to the Distribution of Test Items Among Pupils." *Educational and Psychological Measurement* 18 (1958): 325-29.

Kleinke, D.J. "A Linear-Prediction Approach to Developing Test Norms Based on Matrix-Sampling." *Educational and Psychological Measurement* 32 (1972): 75-84.

Kleinke, D.J. "Further Studies of Linear Prediction Following Matrix Sampling." Paper presented at Annual Meeting of the National Council on Measurement in Education, New Orleans, 1973.

Lord, F.M. "Estimating Norms by Item-Sampling." *Educational and Psychological Measurement* 22 (1962): 259-67.

Lord, F.M. "Use of True-Score Theory to Predict Moments of Univariate and Bivariate Observed-Score Distributions." *Psychometrika* 25 (1960): 325-42.

Lord, F.M. and Novick, M.R. *Statistical Theories of Mental Test Scores*. Reading, Mass.: Addison-Wesley, 1968.

Owens, T.R. and Stufflebeam, D.L. "An Empirical Comparison of

Item Sampling and Examinee Sampling for Estimating Test Norms." *Journal of Educational Measurement* 6 (1969): 75-83.

Plumlee, L.B. "Estimating Means and Standard Deviations from Partial Data—An Empirical Check on Lord's Item Sampling Technique." *Educational and Psychological Measurement* 27 (1964): 623-30.

Shoemaker, D.M. "Allocation of Items and Examinees in Estimating a Norm Distribution by Item-Sampling." *Journal of Educational Measurement* 7 (1970): 123-28. (a)

Shoemaker, D.M. "Item-Examinee Sampling Procedures and Associated Standard Errors in Estimating Test Parameters." *Journal of Educational Measurement* 7 (1970): 255-62. (b)

Shoemaker, D.M. "Test Statistics as a Function of Item Arrangement." *Journal of Experimental Education* 39 (1970): 85-88. (c)

Sirotnik, K. "An Investigation of the Context Effect in Matrix Sampling." *Journal of Educational Measurement* 7 (1970): 199-207.

Wood, R. and Skurnik, L.S. *Item Banking: A Method for Producing School-based Examinations and Nationally Comparable Grades*. London: National Foundation for Educational Research in England and Wales, 1969.

REFERENCES FOR CHAPTER 2

Barcikowski, R.S. "A Monte Carlo Study of Item Sampling (Versus Traditional Sampling) for Norm Construction." *Journal of Educational Measurement* 9 (1972): 209-14.

Gardner, P.L. "Test Length and the Standard Error of Measurement." *Journal of Educational Measurement* 7 (1970):271-73.

Hooke, R. "Symmetric Functions of a Two-way Array." *Annals of Mathematical Statistics* 27 (1956): 55-79. (a)

Hooke, R. "Some Applications of Bipolykays to the Estimation of Variance Components and Their Moments." *Annals of Mathematical Statistics* 27 (1956): 80-98. (b)

Lord, F.M. "Tests of the Same Length Do Have the Same Standard Error of Measurement." *Educational and Psychological Measurement* 19 (1959): 233-39.

Lord, F.M. and Novick, M.R. *Statistical Theories of Mental Test Scores*. Reading, Mass.: Addison-Wesley, 1968.

McMorris, R.F. "Evidence on the Quality of Several Approximations for Commonly Used Measurement Statistics." *Journal of Educational Measurement* 9 (1972): 113-22.

Osburn, H.G. "The Effect of Item Stratification on Errors of Measurement." *Educational and Psychological Measurement* 29 (1969): 295-301.

Rajaratnam, N., Cronbach, L.J. and Gleser, G.C. "Generalizability of Stratified-Parallel Tests." *Psychometrika* 30 (1965): 39-56.

Shoemaker, D.M. and Osburn, H.G. "An Empirical Study of Generalizability Coefficients for Unmatched Data." *British Journal of Mathematical and Statistical Psychology* 21 (1968): 239-46.

Shoemaker, D.M. *A Note on Allocating Items to Subtests in Multiple*

Matrix Sampling. Southwest Regional Laboratory for Educational Research and Development Technical Memorandum, 1972.

Shoemaker, D.M. "Standard Errors of Estimate in Item-Examinee Sampling as a Function of Test Reliability, Variations in Item Difficulty Indices and Degree of Skewness in the Normative Distribution." *Educational and Psychological Measurement* 32 (1972): 705-14.

REFERENCES FOR CHAPTER 3

Bennett, J.H. "Partitions in More than One Dimension." *Journal of the Royal Statistical Society* 18 (1956, Series B): 104-12.

Brandenburg, D.C. and Forsyth, R.A. "Approximating Standardized Achievement Test Norms with a Theoretical Model." Paper presented at the Annual Convention of the National Council on Measurement in Education, New Orleans, 1973. (a)

Brandenburg, D.C. and Forsyth, R.A. "The Use of Multiple Matrix Sampling to Approximate Norm Distributions: An Empirical Comparison of Two Models." Paper presented at the Annual Convention of the National Council on Measurement in Education, New Orleans, 1973. (b)

Carney, E.J. "Relationship of Generalized Polykays to Unrestricted Sums for Balanced Complete Finite Populations." *Annals of Mathematical Statistics* 39 (1968): 643-56.

Collins, J.R. "Jackknifing Generalizability." Ph.D. dissertation. University of Colorado, Boulder, Colorado, 1970.

Cronbach, L.J., Gleser, G.C., Nanda, H. and Rajaratnam, N. *The Dependability of Behavioral Measurements: Theory of Generalizability for Scores and Profiles*. New York: Wiley, 1972.

Dayhoff, E. "On the Equivalence of Polykays of the Second Degree and Sigmas." *Annals of Mathematical Statistics* 35 (1964): 1663-72.

Dayhoff, E. "Generalized Polykays, An Extension of Simple Polykays and Bipolykays." *Annals of Mathematical Statistics* 37 (1966): 226-41.

Elderton, W.P. *Frequency Curves and Correlation*. (Third Edition) Cambridge: Cambridge University Press, 1938.

Elderton, W.P. and Johnson, N.L. *Systems of Frequency Curves*. Cambridge: Cambridge University Press, 1969.

Hill, R.K. "An Alternative Model for Estimating Total-Test Score Distributions Following Item Sampling." Ph.D. dissertation, Syracuse University, 1972.

Hooke, R. *The Estimation of Polykays in the Analysis of Variance*. Princeton University: Statistical Research Group, Memorandum Report No. 56, 1954.

Hooke, R. "Symmetric Functions of a Two-way Array." *Annals of Mathematical Statistics* 27 (1956): 55-79. (a)

Hooke, R. "Some Applications of Bipolykays to the Estimation of Variance Components and Their Moments." *Annals of Mathematical Statistics* 27 (1956): 80-98. (b)

Jones, H.L. "The Jackknife Method." In *Proceedings of the IBM*

Scientific Computing Symposium on Statistics: October 21-23, 1963. White Plains, New York: IBM Data Processing Division, 1965.

Keats, J.A. and Lord, F.M. "A Theoretical Distribution for Mental Test Scores." *Psychometrika* 27 (1962): 59-72.

Kendall, M.G. *The Advanced Theory of Statistics*. Volume I (Fifth Edition) New York: Hafner, 1952.

Kleinke, D.J. "A Linear-Prediction Approach to Developing Test Norms Based on Matrix-Sampling." *Educational and Psychological Measurement* 32 (1972): 75-84.

Knapp, T.R. "An Application of Balanced Incomplete Block Designs to the Estimation of Test Norms." *Educational and Psychological Measurement* 28 (1968): 265-72.

Lord, F.M. "Use of True-Score Theory to Predict Moments of Univariate and Bivariate Observed-Score Distributions." *Psychometrika* 25 (1960): 325-42.

Lord, F.M. "Estimating Norms by Item Sampling." *Educational and Psychological Measurement* 22 (1962): 259-67.

Lord, F.M. "Estimating True-Score Distributions in Psychological Testing (An Empirical Bayes Estimation Problem)." *Psychometrika* 34 (1969): 259-99.

Lord, F.M. and Novick, M.R. *Statistical Theories of Mental Test Scores*. Reading, Mass.: Addison-Wesley, 1968.

Miller, R.G. "A Trustworthy Jackknife." *Annals of Mathematical Statistics* 35 (1964): 1594-1605.

Mosteller, F. "The Jackknife." *Review of the International Statistical Institute* 39 (1971): 363-68.

Mosteller, F. and Tukey, J.S. "Data Analysis, Including Statistics." In Lindzey, G. and Aronson, E. (eds.) *The Handbook of Social Psychology*. (Second Edition) Reading, Mass.: Addison-Wesley, 1968.

Myers, J.L. *Fundamentals of Experimental Design*. Boston: Allyn and Bacon, 1966.

Pandey, T.N. and Shoemaker, D.M. *Estimating Moments of Universe Scores and Associated Standard Errors in Multiple Matrix Sampling for All Item-Scoring Procedures*. Southwest Regional Laboratory for Educational Research and Development Technical Memorandum, 1973.

Pearson, K. "Mathematical Contributions to the Theory of Evolution: Supplement to a Memoir on Skew Variation." *Philosophical Transactions of the Royal Society* 197 (Series A) (1901): 443-59.

Pearson, K. "Contributions to the Mathematical Theory of Evolution: Skew Variation in Homogeneous Material." *Philosophical Transactions of the Royal Society* 186 (Series A) (1895): 343-414.

Quenouille, M.H. "Notes on Bias in Estimation." *Biometrika* 43 (1956): 353-60.

Ray, W.S., Hundleby, J.D. and Goldstein, D.A. "Test Skewness and Kurtosis as Functions of Item Parameters." *Psychometrika* 27 (1962): 39-47.

Shoemaker, D.M. "A Note on Allocating Items to Subtests in Multiple Matrix Sampling and Approximating Standard Errors of Estimate with the Jackknife." *Journal of Educational Measurement* 1973 (In Press).

Shoemaker, D.M. "Allocation of Items and Examinees in Estimating a Norm Distribution by Item-Sampling." *Journal of Educational Measurement* 7 (1970): 123-28.

Shoemaker, D.M. *A Preliminary Investigation of the Theoretical Sampling Distributions of Pooled Estimates of Moments in Multiple Matrix Sampling.* Southwest Regional Laboratory for Educational Research and Development Technical Memorandum, 1973.

Shoemaker, D.M. and Okada, M. *The Communication Skills Program and Concomitant Spelling Proficiency.* Southwest Regional Laboratory for Educational Research and Development Technical Note, 1970.

Sirotnik, K. *Matrix Sampling Estimation in Educational Evaluation: Using the Standard Error of the Mean.* Technical Paper No. 10, The Instructional Objectives Exchange, Los Angeles, Calif., 1973.

Sirotnik, K. "An Analysis of Variance Framework for Matrix Sampling." *Educational and Psychological Measurement* 30 (1970): 891-908.

Winer, B.J. *Statistical Principles in Experimental Design.* (Second Edition) New York: McGraw-Hill, 1971.

REFERENCES FOR CHAPTER 4

Aitchison, J. and Brown, J.A.C. *The Lognormal Distribution.* Cambridge: Cambridge University Press, 1957.

Cook, D.L. and Stufflebeam, D.L. "Estimating Test Norms from Variable Size Item and Examinee Samples." *Educational and Psychological Measurement* 27 (1967): 601-10.

Hooke, R. "Symmetric Functions of a Two-way Array." *Annals of Mathematical Statistics* 27 (1956): 55-79.

Lord, F.M. "Estimating Norms by Item-Sampling." *Educational and Psychological Measurement* 22 (1962): 259-67.

Lord, F.M. "Use of True-Score Theory to Predict Moments of Univariate and Bivariate Observed-Score Distributions." *Psychometrika* 25 (1960): 325-42.

Naylor, T.H., Balintfy, J.L., Burdick, D.S. and Chu, K. *Computer Simulation Techniques.* New York: Wiley, 1966.

Plumlee, L.B. "Estimating Means and Standard Deviations from Partial Data—An Empirical Check on Lord's Item Sampling Technique." *Educational and Psychological Measurement* 24 (1964): 623-30.

Shoemaker, D.M. *A Note on Allocating Items to Subtests in Multiple Matrix Sampling.* Southwest Regional Laboratory for Educational Research and Development Technical Memorandum, 1972.

Shoemaker, D.M. and Osburn, H.G. "A Simulation Model for Achievement Testing." *Educational and Psychological Measurement* 30 (1970): 267-72.

Tucker, L.R. "A Note on the Estimation of Test Reliability by the Kuder-Richardson Formula 20." *Psychometrika* 14 (1949): 117-19.

Wilks, S.S. *Mathematical Statistics.* New York: Wiley, 1962.

REFERENCES FOR CHAPTER 5

Cochran, W.G. "The Comparison of Percentages in Matched Samples." *Biometrika* 37 (1950): 256-66.

Cornfield, J. and Tukey, J.W. "Average Values of Mean Squares in Factorials." *Annals of Mathematical Statistics* 27 (1956): 907-49.

D'Agostino, R.B. "A Second Look at Analysis of Variance on Dichotomous Data." *Journal of Educational Measurement* 8 (1971): 327-33.

Dixon, W.J. (ed.) *BMD Biomedical Computer Programs.* University of California Press, 1968.

Dyke, G.V. and Patterson, H.D. "Analysis of Factorial Arrangements When the Data Are Proportions." *Biometrics* 8 (1952): 1-12.

Gart, J.J. and Zweifel, J.R. "On the Bias of Various Estimators of the Logit and Its Variance with Applications to Quantal Bioassay." *Biometrika* 54 (1967): 181-87.

Hooke, R. "Symmetric Functions of a Two-way Array." *Annals of Mathematical Statistics* 27 (1956): 55-79.

Kirk, R.E. *Experimental Design: Procedures for the Behavioral Sciences.* Belmont, Calif.: Brooks/Cole Publishing Company, 1968.

Light, R.J. and Margolin, B.H. "An Analysis of Variance for Categorical Data." *Journal of the American Statistical Association* 66 (1971): 534-44.

Lunney, G.H. "Using Analysis of Variance with a Dichotomous Dependent Variable: An Empirical Study." *Journal of Educational Measurement* 7 (1970): 263-69.

Satterthwaite, F.E. "An Approximate Distribution of Estimates of Variance Components." *Biometrics Bulletin* 2 (1946): 110-14.

Shoemaker, D.M. "A Note on Allocating Items to Subtests in Multiple Matrix Sampling and Approximating Standard Errors of Estimate with the Jackknife." *Journal of Educational Measurement* 9 (1973), (In Press).

Siegel, S. *Nonparametric Statistics for the Behavioral Sciences.* New York: McGraw-Hill, 1956.

Sirotnik, K. *Some Notes on the Estimation Formulas in Matrix Sampling.* Technical Paper No. 9, The Instructional Objectives Exchange, Los Angeles, Calif., 1973.

Snedecor, G.W. and Cochran, W.G. *Statistical Methods.* Ames, Iowa: Iowa State University Press, 1967.

Winer, B.J. *Statistical Principles in Experimental Design.* (Second Edition) New York: McGraw-Hill, 1971.

Winsor, C.P. "Factorial Analysis of a Multiple Dichotomy." *Human Biology* 20 (1948): 195-204.

REFERENCES FOR CHAPTER 6

Binder, A. "Considerations of the Place of Assumptions in Correlational Analysis." *American Psychologist* 14 (1959): 504-10.

Burton, N. and Remer, R. *Item Sampling in Questionnaires: Rate of Return and Context Effects.* University of Colorado, Laboratory of Educational Research Report No. 57, 1972.

Carroll, J.B. "The Nature of the Data, or How to Choose a Correlation Coefficient." *Psychometrika* 26 (1961): 347-72.

Champion, D.J. and Sear, A.M. "Questionnaire Response Rate: A Methodological Analysis." *Social Forces* 47 (1969): 335-39.

Christian, F.T. *Plan for Educational Assessment in Florida: Final Report.* Florida State Department of Education, Division of Elementary and Secondary Education, Bureau of Research, 1971.

Knapp, T.R. "An Application of Balanced Incomplete Block Designs to the Estimation of Test Norms." *Educational and Psychological Measurement* 28 (1968): 265-72.

Loadman, W.E. *An Inquiry Concerning the Use of Item Sampling as a Method to Reduce Testing Time.* Indiana University, Educational Research and Evaluation Laboratory, Report No. 5, 1972.

Lord, F.M. "Use of True-Score Theory to Predict Moments of Univariate and Bivariate Observed-Score Distributions." *Psychometrika* 25 (1960): 325-42.

Lord, F.M. "Test Norms and Sampling Theory." *Journal of Experimental Education* 27 (1959): 247-63.

Olkin, I. and Pratt, R.W. "Unbiased Estimation of Certain Correlation Coefficients." *Annals of Mathematical Statistics* 29 (1958): 201-11.

Osburn, H.G. "A Note on Design of Test Experiments." *Educational and Psychological Measurement* 27 (1967): 797-802.

Peterson, D.F. and Anderson, D.H. "Closing the Communications Gap with Item Sampling." Paper presented at the Annual Meeting of the American Educational Research Association, 1971.

Piper, R.M. "Multiple Matrix Sampling in the Evaluation of a Music Program." Paper presented at the Annual Meeting of the National Council on Measurement in Education, 1972.

Pugh, R.C. "Empirical Evidence on the Application of Lord's Sampling Technique to Likert Items." *Journal of Experimental Education* 39 (1971): 54-56.

Ray, W.S., Hundleby, J.D. and Goldstein, D.A. "Test Skewness and Kurtosis as Functions of Item Parameters." *Psychometrika* 27 (1962): 34-47.

Riles, W. *Final Report of the Test Development Project of the Miller Mathematics Improvement Program.* Sacramento, Calif.: California State Department of Education, 1971.

Ruben, H. "Some New Results on the Distribution of the Sample Correlation Coefficient." *Journal of the Royal Statistical Society* Series B (1966): 513-25.

Shoemaker, D.M. "An Application of Item-Examinee Sampling to Scaling Attitudes." *Journal of Educational Measurement* 8 (1971): 279-82.

Shoemaker, D.M. "Evaluating the Effectiveness of Competing Instructional Programs." *Educational Researcher* 1 (1972): 5-8.

Sletto, R.F. "Pretesting of Questionnaires." *American Sociological Review* 15 (1950): 24-35.

Tardif, R.F. *Report of the November 1971 SCIMA Test Results.* California State Department of Education, Office of Program Evaluation, 1971.

Timm, N.H. "The Estimation of Variance-Covariance and Correlation Matrices from Incomplete Data." *Psychometrika* 35 (1970): 417-37.

REFERENCES FOR APPENDIX A

Aitchison, J. and Brown, J.A.C. *The Lognormal Distribution.* Cambridge: Cambridge University Press, 1957.

Elderton, W.P. and Johnson, N.L. *Systems of Frequency Curves.* New York: Cambridge University Press, 1969.

Hill, R.K. "An Alternative Model for Estimating Total-Test Score Distributions Following Item Sampling." Ph.D. dissertation: Syracuse University, 1972.

Hooke, R. *The Estimation of Polykays in the Analysis of Variance.* Princeton University: Statistical Research Group, Memorandum Report No. 56, 1954.

Hooke, R. "Symmetric Functions of a Two-way Array." *Annals of Mathematical Statistics* 27 (1956): 55-79.

Lord, F.M. and Novick, M.R. *Statistical Theories of Mental Test Scores.* Reading, Mass.: Addison-Wesley, 1968.

Myers, J.L. *Fundamentals of Experimental Design.* Boston: Allyn and Bacon, 1966.

Pandey, T.N. and Shoemaker, D.M. *Estimating Moments of Universe Scores and Associated Standard Errors in Multiple Matrix Sampling for All Item-Scoring Procedures.* Southwest Regional Laboratory for Educational Research and Development Technical Memorandum, 1973.

Shoemaker, D.M. *A General Procedure for Approximating Standard Errors of Estimate in Multiple Matrix Sampling.* Southwest Regional Laboratory for Educational Research and Development Technical Memorandum, 1972.

Winer, B.F. *Statistical Principles in Experimental Design.* (Second Edition) New York: McGraw-Hill, 1971.

Bibliography

Barcikowski, R.S. "A Computer Program for Randomly Selecting Test Items from an Item Population." *Educational and Psychological Measurement* 32 (1972): 795-98.

Barcikowski, R.S. "A Monte Carlo Study of Item Sampling (Versus Traditional Sampling) for Norm Construction." *Journal of Educational Measurement* 9 (1972): 209-14.

Burton, N. and Remer, R. *Item Sampling in Questionnaires: Rate of Return and Context Effects.* Laboratory of Educational Research, University of Colorado, Report No. 57, 1972.

Cahen, L.S. "An Interim Report on the National Longitudinal Study of Mathematical Abilities." *The Mathematics Teacher* 58 (1965): 522-26.

Cahen, L.S., Romberg, T.A. and Zwirner, W. "The Estimation of Mean Achievement Scores for Schools by the Item-Sampling Technique." *Educational and Psychological Measurement* 30 (1970): 41-60.

Cook, D.L. and Stufflebeam, D.L. "Estimating Test Norms from Variable Size Item and Examinee Samples." *Educational and Psychological Measurement* 27 (1967): 601-10.

Cook, D.L. and Stufflebeam, D.L. "Estimating Test Norms from Variable Size Item and Examinee Samples." *Journal of Educational Measurement* 4 (1967): 27-33.

Cronbach, L.J. "Course Improvement Through Evaluation." *Teachers College Record* 64 (1963): 672-83.

Cronbach, L.J., Rajaratnam, N. and Gleser, G.C. "Theory of Generalizability: A Liberalization of Reliability Theory." *British Journal of Statistical Psychology* 16 (1963): 137-63.

Cronbach, L.J., Schönemann, P. and McKie, D. "Alpha Coefficients for Stratified-Parallel Tests." *Educational and Psychological Measurement* 25 (1965): 291-312.

Das, R.S. "Some Models for Assessment of Intelligence and Scholastic Attainment." *Journal of Psychological Research* 11 (1967): 77-90.

Denenberg, V.H. and Besco, R.O. "A Method of Obtaining a Common Rank Order Scale from Incomplete Data." *Personnel Psychology* 14 (1961): 207-10.

Gorth, W. and Grayson, A. "A Program to Compose and Print Tests for Instructional Testing Using Item Sampling." *Educational and Psychological Measurement* 29 (1969): 173-74.

Gorth, W., Grayson, A., Popejoy, L. and Stroud, T. "A Tape-based Data Bank for Educational Research or Instructional Testing Using Longitudinal Item Sampling." *Educational and Psychological Measurement* 29 (1969): 175-77.

Gorth, W., Grayson, A. and Stroud, T. "A Computer Program to Tabulate and Plot Achievement Profiles of Longitudinal Achievement Testing Using Item Sampling." *Educational and Psychological Measurement* 29 (1969): 179-80.

Gorth, W., Grayson, A. and Lindeman, R. "A Computer Program to Evaluate Item Performance by Internal and External Criteria in a Longitudinal Testing Program Using Item Sampling." *Educational and Psychological Measurement* 29 (1969): 181-83.

Hambleton, R.K., Rovinelli, R. and Gorth, W.P. "Efficiency of Various Item-Examinee Sampling Designs for Estimating Test Parameters." *Proceedings*, 79th Annual Convention, American Psychological Association, 1971.

Hooke, R. "Some Applications of Bipolykays to the Estimation of Variance Components and Their Moments." *Annals of Mathematical Statistics* 27 (1956): 80-98.

Hooke, R. "Symmetric Functions of a Two-way Array." *Annals of Mathematical Statistics* 27 (1956): 55-79.

Husek, T.R. and Sirotnik, K. "Matrix Sampling." *UCLA Evaluation Comment* 1 (1968): 1-4.

Johnson, M.C. and Lord, F.M. "An Empirical Study of the Stability of a Group Mean in Relation to the Distribution of Test Items Among Pupils." *Educational and Psychological Measurement* 18 (1958): 325-29.

Keats, J.A. and Lord, F.M. "A Theoretical Distribution for Mental Test Scores." *Psychometrika* 27 (1962): 59-72.

Kleinke, D.J. "A Program for Generating a Negative Hypergeometric Distribution for Test Score Data." *Educational and Psychological Measurement* 30 (1970): 745-46.

Kleinke, D.J. "A Linear-Prediction Approach to Developing Test Norms Based on Matrix-Sampling." *Educational and Psychological Measurement* 32 (1972): 75-84.

Knapp, T.R. "An Application of Balanced Incomplete Block Designs to the Estimation of Test Norms." *Educational and Psychological Measurement* 28 (1968): 265-72.

Loadman, W.E. *An Inquiry Concerning the Use of Item Sampling as a Method to Reduce Testing Time.* Educational Research and Evaluation Laboratory, Indiana University, Report No. 5, 1972.

Lord, F.M. "Sampling Fluctuations Resulting from the Sampling of Test Items." *Psychometrika* 20 (1955): 1-23.

Lord, F.M. "An Approach to Mental Test Theory." *Psychometrika* 24 (1959): 283-302.

Lord, F.M. "Test Norms and Sampling Theory." *Journal of Experimental Education* 27 (1959): 247-63.

Lord, F.M. "Statistical Inference about True Scores." *Psychometrika* 24 (1959): 1-17.

Lord, F.M. "Use of True-Score Theory to Predict Moments of Univariate and Bivariate Observed-Score Distributions." *Psychometrika* 25 (1960): 325-42.

Lord, F.M. "Estimating Norms by Item-Sampling." *Educational and Psychological Measurement* 22 (1962): 259-67.

Lord, F.M. and Novick, M.R. *Statistical Theories of Mental Test Scores*. Reading, Mass.: Addison-Wesley, 1968. (Chapter 11: "Item Sampling in Test Theory and in Research Design": Chapter 23: "Binomial Error Models.")

McCormick, E.J. and Bachus, J.A. "Paired Comparison Ratings: I. The Effect on Ratings of Reductions in the Number of Pairs." *Journal of Applied Psychology* 36 (1952): 123-27.

McCormick, E.J. and Roberts, W.K. "Paired Comparison Ratings: II. The Reliability of Ratings Based on Partial Pairings." *Journal of Applied Psychology* 36 (1952): 188-92.

Osburn, H.G. "A Note on Design of Test Experiments." *Educational and Psychological Measurement* 27 (1967): 297-302.

Osburn, H.G. "Item Sampling for Achievement Testing." *Educational and Psychological Measurement* 28 (1968): 95-104.

Osburn, H.G. "The Effect of Item Stratification on Errors of Measurement." *Educational and Psychological Measurement* 29 (1969): 295-301.

Owens, T.R. and Stufflebeam, D.L. "An Experimental Comparison of Item Sampling and Examinee Sampling for Estimating Test Norms." *Journal of Educational Measurement* 6 (1969): 75-83.

Piper, R.M. and Shoemaker, D.M. "Formative Evaluation of a Kindergarten Music Program Based on Behavioral Objectives. *Journal of Research in Music Education* 21 (1973); (In Press).

Plumlee, L.B. "Estimating Means and Standard Deviations from Partial Data—An Empirical Check on Lord's Item Sampling Technique." *Educational and Psychological Measurement* 24 (1964): 623-30.

Pugh, R.C. "Empirical Evidence on the Application of Lord's Sampling Technique to Likert Items." *Journal of Experimental Education* 39 (1971): 54-56.

Rajaratnam, N., Cronbach, L.J. and Gleser, G.C. "Generalizability of Stratified-Parallel Tests." *Psychometrika* 30 (1965): 39-56.

Shoemaker, D.M. and Osburn, H.G. "An Empirical Study of Generalizability Coefficients for Unmatched Data." *British Journal of Mathematical and Statistical Psychology* 21 (1968): 239-46.

Shoemaker, D.M. and Osburn, H.G. "Computer-Aided Item Sampling for Achievement Testing: A Description of a Computer Program Implementing the Universe Defined Test Concept. *Educational and Psychological Measurement* 29 (1969): 165-72.

Shoemaker, D.M. "Allocation of Items and Examinees in Estimating a Norm Distribution by Item-Sampling." *Journal of Educational Measurement* 7 (1970): 123-28.

Shoemaker, D.M. "Test Statistics as a Function of Item Arrangement." *Journal of Experimental Education* 39 (1970): 85-88.

Shoemaker, D.M. "Item-Examinee Sampling Procedures and Associated Standard Errors in Estimating Test Parameters." *Journal of Educational Measurement* 7 (1970): 255-62.

Shoemaker, D.M. "Further Results on the Standard Errors of Estimate Associated with Item-Examinee Sampling Procedures." *Journal of Educational Measurement* 8 (1971): 215-20.

Shoemaker, D.M. "An Application of Item-Examinee Sampling to Scaling Attitudes." *Journal of Educational Measurement* 8 (1971): 279-82.

Shoemaker, D.M. "Evaluating the Effectiveness of Competing Instructional Programs." *Educational Researcher* 1 (1972): 5-8.

Shoemaker, D.M. "Standard Errors of Estimate in Item-Examinee Sampling as a Function of Test Reliability, Variation in Item Difficulty Indices and Degree of Skewness in the Normative Distribution." *Educational and Psychological Measurement* 32 (1972): 705-14.

Shoemaker, D.M. "Improving Criterion-Referenced Measurement." *Journal of Special Education* 6 (1972): 315–23.

Shoemaker, D.M. "A Note on Allocating Items to Subtests in Multiple Matrix Sampling and Approximating Standard Errors of Estimate with the Jackknife." *Journal of Educational Measurement* 9 (1973), (In Press).

Sirotnik, K. "An Investigation of the Context Effect in Matrix Sampling." *Journal of Educational Measurement* 7 (1970): 199-207.

Sirotnik, K. "An Analysis of Variance Framework for Matrix Sampling." *Educational and Psychological Measurement* 30 (1970): 891-908.

Wood, R. and Skurnik, L.S. *Item Banking: A Method for Producing School-based Examinations and Nationally Comparable Grades.* London: National Foundation for Educational Research in England and Wales, 1969.

Index

About the Author

Dr. David M. Shoemaker is a Senior Member of the Professional Staff at the Southwest Regional Laboratory for Educational Research and Development in Los Alamitos, California. He received the A.B. in Psychology from Indiana University, the M.A. in Industrial Psychology from Bowling Green State University in Ohio, and the Ph.D. in Industrial Psychology from the University of Houston. His research interests concern the evaluation of instructional programs, large-scale assessment, and computer simulation, and he is the author or coauthor of more than 25 scientific articles and reports. He was an Assistant Professor in the Department of Psychology at Oklahoma State University before joining the staff of the Southwest Regional Laboratory and has served subsequently as Consultant to the State of California in the design of state-wide assessment programs. Dr. Shoemaker is a member of the American Psychological Association.